Lean Culture for the Construction Industry

Building Responsible and Committed Project Teams

Lean Culture for the Construction Industry

Building Responsible and Committed Project Teams

GARY SANTORELLA

CRC Press
Taylor & Francis Group
Boca Raton   London   New York

CRC Press is an imprint of the
Taylor & Francis Group, an **informa** business

A PRODUCTIVITY PRESS BOOK

Productivity Press
Taylor & Francis Group
270 Madison Avenue
New York, NY 10016

© 2011 by Taylor and Francis Group, LLC
Productivity Press is an imprint of Taylor & Francis Group, an Informa business

No claim to original U.S. Government works

Printed in the United States of America on acid-free paper
10 9 8 7 6 5 4 3 2 1

International Standard Book Number: 978-1-4398-3508-1 (Hardback)

Library of Congress Cataloging-in-Publication Data

Santorella, Gary.
    Lean culture for the construction industry : building responsible and committed project teams / Gary Santorella.
        p. cm.
    Includes bibliographical references and index.
    ISBN 978-1-4398-3508-1 (hardcover : alk. paper)
    1. Construction industry--Personnel management. 2. Construction industry--Management. 3. Lean manufacturing. I. Title.

HD9715.A2S345 2011
624.068'4--dc22   2010045372

Visit the Taylor & Francis Web site at
http://www.taylorandfrancis.com

and the Productivity Press Web site at
http://www.productivitypress.com

Anyone can become angry—that is easy. But to be angry with

the right person, to the right degree, at the right time, for the

right purpose, and in the right way—that is not easy.

—**Aristotle,** *The Nichomachean Ethics*

It is with the heart that one sees rightly; what

is essential is invisible to the eye.

—**Antoine De Saint-Exupery,** *The Little Prince*

Much unhappiness has come into the world because

of bewilderment and things left unsaid.

—**Fyodor Dostoevsky**

Contents

Foreword

The construction industry is exceptional. Our products have stood the test of time and stand as lasting reminders of the collective work of hundreds of architects, engineers, and craftsmen whose combined efforts build the very structures we admire, the places we work, and the facilities that ensure our health and well-being. Combining brains and brawn, this small group of individuals makes up about 12% of the GDP and transforms dreams into reality.

Unfortunately, these feats of construction have not been easy. Territorial in nature, each discipline, each company, each trade, and even each individual function of the construction process seeks to maximize its own results, even if they come at the expense of the overall project objective or other members of the team. In the end, the project is completed but the cost of completion both financially and emotionally is often too high to recover.

While much attention is given to the financial cost, little has been done to address the emotional impact on the project participants, the underlying root cause of the financial downfall and one of the major factors causing a widening concern in the construction industry.

The construction industry is faced with an ever-growing shortage of an effective, skilled workforce. Marked by a propensity for conflict at every level, it is very true that "buildings leak at the intersection of contracts." Whether that contractual relationship is between the owner and the general contractor, the general contractor and the subcontractor, or a supervisor and his or her direct reports, these conflicts are having a direct impact on the financial results of the project and driving the best and brightest from the construction industry.

But the good news is that over the past several years our industry has sought change—revisions to the way we approach work, the way we collaborate, every aspect of the way we treat one another. The growth of design build and other forms of integrated project delivery is evidence of how construction is working to openly engage all participants in the building process. We are finally getting it: construction is more than a zero-sum game. Marketing, engineering, planning, architecture, and production strategies must be executed in harmony.

We know all too well that attempts at change, especially ones of this magnitude, are difficult at best, especially in our industry, where when placed under stress, we have a tendency to revert to traditional ways of doing things, seeking conflict rather than collaboration, excuses rather than solutions, and blame rather than results. This cannot be business as usual or construction will never really progress.

"SEEK FIRST TO UNDERSTAND BEFORE YOU SEEK TO BE UNDERSTOOD"

Transformation needed in the construction industry will require old fractional job descriptions to change, maybe even blur, as stakeholders focus their energies on the best interest of the project, not themselves, their trade, or their organization. We must accept responsibility for every project participant's success. Construction is not a zero-sum game; someone doesn't have to lose for someone to win.

People have used the word *accountability* in business for years. Books have been devoted to the subject. Our industry brandishes it like a weapon—we expect, even demand accountability. If something isn't going well, someone must pay!

While I believe we must all hold ourselves accountable to one another, there is one aspect of accountability that has always caused me concern, and that is that it is a "lagging indicator" of something that has already happened. To me accountability says, "Let's wait and see what happens and then we will decide who gets the blame or credit."

Instead I like the word *responsibility*. Up front, right now, before anything has even happened, we are responsible for and to one another for the results that follow. We will not let one another fail. Instead, each project participant is looking out for one another, all in harmony with the overall project objectives of the client, the project owner. While more the exception than the rule, when it is applied in our industry as it is with one of our long-term clients, we see production increased by as much as 40%, customer satisfaction at its highest level in our organization, repeat business a given, and a happy and effective workforce that feels engaged and takes pride in the results we provide!

This is the reason I told Gary I would write this foreword. Our industry does need change.

Over the years, Gary has been instrumental in the evolution of our organization, helping us through everything from strategic planning to personal counseling. His engagement has helped us be a better organization. By working with our teams on most of the principles outlined in this book, we have seen our company grow tenfold, our customer satisfaction improve dramatically, and our geographic influence expand.

I'm sure that you will find Gary's approach to our industry fresh, enlightening, and beneficial to your organization. He will challenge your conventional thinking and provide you with practical applications that you will be able to implement quickly in your own organization, regardless of what position you hold.

Tom Sorley, CEO
Rosendin Electric

Acknowledgments

At the risk of inadvertently leaving someone out, I wanted to thank all the wonderful and dedicated men and women who took time out of their pressure-packed schedules to patiently answer my ridiculous questions and teach me about the intricacies of their complicated profession. Without the following people (and countless others), this book simply could not have been written: Tom Sorley, Paul Pettersen, Dick Dorais, John DiCiurcio, Frank O'Connor, Kenneth Leach, Scott Holbrook, Charlie Murphy, Craig Holt, Craig Bjorkman, John DeRuiter, Larry Beltramo, Jim Hawk, Jeff Loyall, Jackie Costigan, Eric Wildt, Dwayne Goddard, Joe Lucarelli, David Klopp, Erik Walker, Tom Turner, Ron Rudolph, Rob Stein, Elmond Wan, Larry Atwater, David Tsao, Ed Cadena, Diane O'Carroll, Pat Di Filippo, Rod Michalka, Rich Bach, Bruce Ruthoen, Rodney Pope, George Zettle, Dan Gemme, Dan Kavanaugh, Mary Marshall, John Weaver, Terry Shugrue, Jim Goldman, Jack Beaudoin, Ken Schroeder, LeRoy King, Dave Holland, J.P. Bol, Rick Shandrew, Steve Scates, John Koester, Randy Hirotsu, John Brown, Saptarshi Desei, Joel McRonald, Michael Shook, Jim Delaney, Emil Konrath, Ron and Cindy McMackin, Bob Berenguer, Willie Micene, Steve Foxworthy, Mike Roberts, Frank DaiZovi, Terry Richards, Harry Smith, Steve Annese, Thang Do, Char Kaufer, Kathleen McCartney, Bill Mazzetti, Dennis Newman, Scott Miller, Dennis Lichty, Cooper Mitchell-Rekrut, Dr. Stephen Misouich, Michael Rekrut, Dr. Stephen Misovich, Geoff Ross, and Ray Munger.

And a very special thanks to Lisa Vere, Bruce Wexler, Gus Sestrap, E.J. Saucier, Tom Gerlach, T.J. Lyons, Stacy Sakellarides, JoAnn Ferrante, Laura Reyes, and my parents for their invaluable feedback, thoughtful tutelage, and undying (often unwarranted) support and encouragement.

Introduction

> There are two kinds of people in the world: Those who are fools because they think that they are wise; and those who are wise because they know that they are fools.
>
> —Socrates

My infatuation with the construction industry began with a phone call in June 1996. My wife, then a project manager for a general contractor, received an urgent message from her operations manager: "We need someone to do team building," he said. "Doesn't your husband do (expletive deleted) like this?" "Yes," she deadpanned, "My husband does (expletive deleted) like this." And I've been doing it ever since—and loving every minute of it.

What's not to love? It is, after all, a noble enterprise, one in which we alter our world by creating structures that make our harsh surroundings more hospitable. Beyond the practical, the efforts of construction professionals—from highbrow architects to sweat-of-the-brow laborers—determine the greatness of a city and a civilization. What would New York, Chicago, London, Paris, Moscow, or Beijing be without their unique churches, temples, houses, apartments, skyscrapers, bridges, roads, and infrastructures that define them?

I love construction people unreservedly, particularly those in the thankless role of middle management. Yes, they can be stubborn and even arrogant at times. But oddly enough, these same qualities are what make them so endearing to work with, they consistently go about their business with unusual passion, pride, and the intent to do the best job possible. Decently paid but overworked and underappreciated, these managers perform a juggling act beyond comprehension. They routinely make sense of, organize, and build off of literally thousands of pages of contract documents and conceptual drawings—all of which are constantly being altered and amended via a seemingly endless stream of emails and bridging documents. Not only are they charged with physically getting the project built, but they are called upon to balance the often adversarial needs of designers, owners, lawyers, government bureaucrats, and inspectors—all the while trying to buy out materials and services, conceptualize the flow of the work, and determine the best means and methods possible to reach the goals set forth by their own companies (i.e., find a way to make a meager profit).

On top of this, the people they oversee—those who do the actual work—are a wildly varied lot, ranging from the brilliant and talented to the self-serving and slothful. The former are true artists and geniuses at their craft, while a significant number of the latter often have one foot at the job site and the other close to the county jail.

Somehow amid this confluence of competing interests, challenging work-force dynamics, and information overload, operations managers, project executives, project managers, superintendents, general foremen, and foremen bring order to the chaos, provide direction and key decisions, and get the job finished on time, on budget, and to design—regardless of the insanity swirling around them. The fact that they are able to get it right as often as they do is a real testament to their intelligence and intestinal fortitude.

But despite all of their efforts, waste bedevils every construction project. Over the past forty years, productivity for most manufacturing processes in this country has gone up by over 100% (1.77% per year). In this same time period, productivity in the construction industry, as measured by contract value/labor hours, has gone down by a staggering 25% (or an average compound rate of −0.59% per year).* In recent years, advocates of Lean construction have cited numerous causes of waste, including poor planning and scheduling, ineffective methods of procurement and material handling, and haphazard work plans. What is talked about less frequently is the amount of waste that is directly related to ineffective leadership. A great deal of lost productivity and lost revenue is caused by poorly devised organizational structures, unclear roles and responsibilities, unresolved interpersonal conflicts that are allowed to fester, and an overall lack of focus on improving team process—all of which is under the direct purview of project leaders.

That is why I am writing this book—to draw a connection between how construction professionals act as leaders (both positively and negatively) and the subsequent affect their attitude and behavior has on productivity and waste—each and every day. To me, this is the very essence behind most Lean principles—understanding the impacts that the supposed little

---

* From AECbytes viewpoint 4, by Paul Teicholz, PhD, Professor Emeritus, Stanford University.

things have on what, in today's economy, is a very crucial thing indeed—the bottom line.

Simply put, Profit = (Price − Cost) × Volume. Given that the greatest risk factor on any project is manpower costs, anything on the people side of the job that results in delays, reworks, or overtime will lower profits via increased labor costs. And the sad fact is most of these people-generated costs are fully preventable.

As much as I like and respect construction professionals, I have found that they can be their own worst enemies. Most construction managers greatly underestimate how hiccups on the people side can negatively impact the quality and profitability of the product they generate on the technical side—and how they themselves can, inadvertently, be the biggest contributor to these types of problems. Over the years, I've seen even the best managers struggle with the following:

- Failing to conceptualize, create, and distribute an organizational structure that accurately reflects how project flow and communication actually works, resulting in confusion and numerous processing and coordination hiccups.
- Paying too little attention to *how* people are communicating and instead overvaluing people for merely "cranking out" work, thus creating a high volume of uncoordinated people-generated inventory that is often either poorly executed or simply gets lost in the informational sauce.
- Favoring short-term fixes over long-term solutions; often dropping down to do someone's job for them instead of teaching them how to do their job, thereby creating unintended people problems in the form of leadership-assisted incompetence.
- Possessing insufficient understanding about how to influence people; instead, often punishing people for showing initiative or, conversely, rewarding them for doing things that violate standard practices.
- Neglecting to create a forum to identify critical issues and to strategize how to tackle these issues—as a team. They inadvertently reinforce the repetition of poor performance.
- Underestimating the impact that unresolved conflicts and poor role delineation can have on productivity and how these dysfunctions can contribute to issues falling through the cracks, long lead items not being identified, or work put in place without proper owner approval.

- Overvaluing people for their technical prowess and often overlooking their "team killing" behaviors, such as when a technical "star" talks negatively behind a teammate's back, hoards information, publicly tears someone down, or doesn't provide coaching or training—and failing to appreciate how these behaviors can negatively impact profitability.
- Being thing-aware, not self-aware. While they are phenomenal at solving technical problems, they often fail to see how having a poor attitude (pointing fingers when mistakes occur) or their own lack of approachability can prove harmful to project flow.
- Underestimating the need people have for a sense of purpose about what they do.
- Becoming overly enamored with engineering processes and procedures, and thus viewing people merely as objects who satisfy these functions rather than as individuals with distinct needs, concerns, and goals.
- Being adrenaline junkies; they want to go, go, go, when they should really slow down and build a unified plan.
- Being mavericks and individualists, they often unwittingly choose to reinvent the wheel rather than ask for input, help, or advice.

Any of these listed attitudes and behaviors alone can cripple productivity and result in waste and lost profit. In combination, they are truly deadly. Creating a team out of a random bunch of people who just happen to be stuck in a trailer together is a true art, but some construction managers rely on a more prosaic approach—they simply put people with the right technical skills together and call it a team. My hope is that this book will convince them leadership skills that go hand in hand with Lean thinking, an approach that involves far more than just-in-time delivery, Pareto charts, pull planning, and value stream mapping. The art of team building in a Lean world is about the art of culture building, i.e., developing new ways to think about *how* you do what you do more productively and with less waste.

Transitioning from doer to leader and developing this Lean mindset is a particular challenge for construction industry professionals. Most people get into this business because they like the technical side of the work. They enjoy figuring out how to dewater the site or how to get a balky new exterior curtain wall system to work. In many cases, particularly for those who grew up with fathers and grandfathers in the industry, it's literally a part of their DNA. Most construction managers have been lauded, promoted, and given other perks precisely because of their ability to achieve short-term results. That's why most have little trouble writing purchasing requisitions, drawing up schedules, generating manpower reports, writing potential change orders (PCOs), or tracking requests for information (RFIs), but often struggle with how to transmit the knowledge contained within these documents to their people, or find it difficult to foster the communication or organizational structure required to knit the people on their team into a well-functioning whole. Therefore, the same person who has the ability to look out his or her office window and immediately discern that the conduit on the eighth floor has been installed incorrectly, may at the same time be perplexed by his or her staff's lack of urgency or tendency to repeat the same mistakes over and over again.

This book is precisely for these folks. It is intended as an accessible reference that can be easily adapted to just about any job site situation—and something that people in construction won't have to translate from a generic "one size fits all" list of tired managerial tropes at their local bookstore. It offers the tools to cope with the nontechnical side of their job, and the confidence that they can deal with just about any people situation they encounter efficiently and effectively. Most importantly, it explains how such seemingly non-construction-related activities such as developing trust, direct communication, role definition and clarity, and actively engaging with one's team will have a positive impact on a team's efficiency and the bottom line.

But before moving forward, please allow me to explain why someone with my credentials as a cognitive behavioral therapist is qualified to write a book for construction professionals. I could tell you about my fourteen years of experience as a consultant with all sorts of companies in this industry, or my background in the corrugated box industry, but I'd prefer to share one specific story.

Many years ago, I was hired to facilitate a two-day partnering/conflict resolution session between a general contractor (in the role of construction managers [CM]) and various mechanical, electrical, and plumbing (MEP) subcontractors on a $350 million dollar CM/multiple prime public hospital job. The project was plagued by poor coordination and interrupted workflow, which in turn triggered numerous back and forth bouts of accusatory emails and formal letters. What brought things to a head was an incident where the plumbers had deliberately ripped out drywall well beyond what was required in order to access a particular area of the building. Unfortunately, on the first day, I was greeted with a heavy dose of skepticism and wariness as I tried to deal directly with the issues.

After a night of heavy drinking (at the general constructions [GC's] expense) and other eyebrow raising activities that I managed to avoid, the next day, the group dragged themselves in—predictably hung over, curiously bruised (there are some things that I really don't want to know), and more than a little embarrassed. Exploiting their guilt and physically compromised condition for all it was worth, and with the aid of several gallons of high-octane coffee, we actually got down to the business of assessing the situation. But unlike the previous day, the comments expressed weren't dripping with sarcasm. As we probed deeper, it became clear how truly frustrated everyone was with this job—and had been since its inception. And what stood out even more was the fact that nobody really knew why. All they knew for certain was that they were sick and tired of repeatedly showing up on site and being told that an area that they were expecting to work in wasn't ready or, conversely, being berated for not "manning up" to unexpected requirements and losing gobbs of money in the process. Of course, they all had plenty of theories as to why this was happening: the damn architect wasn't responding to their RFIs in a timely fashion; they weren't getting adequate direction from the CM; some (unnamed) people didn't care about the job as much as they did; and yes, maybe *they* hadn't been attending the coordination meetings as they should have been.

But the more they talked, the more the true root cause became apparent. And this is where my ignorance of the industry actually became an advantage. As I continued to ask questions about the contract and how the job was set up, I kept noting the quizzical looks that were popping up on the faces around the table. Despite the fact that they were working under a CM/multiple prime contract for nearly two years, most of the people in the room had no idea what it meant to work in such a delivery system. The subcontractors kept treating the job like a regular GC job (waiting for the GC to generate updated work schedules, waiting for the various and sundry reporting requests to trickle down from the GC, waiting for the GC to give them specific daily direction) when the reality was, in this type of delivery system, per contract, it was the subcontractors' responsibility to assume the primary role of project coordination, scheduling day-to-day activities as well as adhering to vital administrative processes. Yes, the CM was to provide a master schedule, but in this delivery system, it wasn't the CM's job to provide specific daily direction. They were to provide clarification, information, and guidance—not for their own plan, but for the one that the *subcontractors* were supposed to generate and follow.

Even though these expectations had been identified in a high-level meeting at the beginning of the job, the subtleties of this delivery system did not trickle down to the most important people—the on-site managers who were expected to run the job in the day to day. So, as is usually the case when people are under stress, they simply reverted to familiar, overlearned behaviors. And in this case, this meant that everyone was treating this job as if it were a traditional GC job. Making matters worse, the CM hadn't been direct about their own frustrations. When they saw things going awry, they took a somewhat passive-aggressive stance. Instead of shutting down the job and recalibrating the process, they dogmatically stuck to the contract and withheld all forms of direction without providing any clarification as to why they were doing so. At other times, they directed their own laborers to do tasks that were in the direct purview of the subcontractors, which was perceived as a negative intrusion by the subcontractors. Quite often, the CM went so far as to simply withhold payment without explanation—particularly when they determined that proper administrative procedures hadn't been followed. Why did the CM act in this manner even though the data suggested that such behaviors weren't helping the situation? The CM had convinced themselves that by acting in an indirect fashion, the subcontractors would somehow magically "get the message" and fall in line. Unfortunately, all this actually did was introduce a tremendous amount of waste into the system.

After a thorough and objective discussion, everyone had the same epiphany: if they didn't take a step back and treat this session as if it were the first day of the project, this is, redefining roles—and expectations of these roles—per the contract and committing to carrying out the implied promises of these roles each day, this job stood no chance of recovery. So that is precisely what we did. To everyone's credit, they were able to let go of the past and formulate a plan to carry the job forward. By the end of the second day, everyone was cautiously optimistic.

That's when I realized I had something to offer these construction professionals. The problems that were infecting this project weren't at all technical or engineering in nature—those kinds of problems always had a solution. It was the lack of role clarity, the unspoken frustrations, the false assumptions about each other's poor performance, the implied accusations of wrongdoing, the angrily withheld exchanges of information, the well-intended manipulations in the heat of action, the breaks in chain of command for expediency sake—these were issues that I knew how to deal with, and they, because of their highly technical backgrounds, didn't.

Over time, I became aware that my years of assessing and working with teams dovetailed well with Lean principles. Let me give you a sense of how this is so. I had developed my own method to assess team functioning, largely based in cognitive behavioral and industrial psychology. For instance, I would evaluate how much time is spent on value-added activity (doing actual work that contributes to the project) versus time wasted due to communication misfires, confusion, or unwanted duplication of service. Through this evaluation, I discovered the factors that impede productivity and produce waste. Using this information, I worked with managers and helped them to think in broader terms about a project. As a result, project management teams are able to direct their energy to areas that really matter. By using the data derived from my work with these teams, I could help them measure their progress and facilitate continuous improvement. In essence, I could help them shift their thinking away from focusing on short-term results, to taking a broader view that would help achieve overall project success.

After doing this type of work for some time, and as Lean construction practices began to take hold, people well-versed with such thinking told me that my methods were aligned with Lean principles. The more I read and talked to people about Lean concepts, the more I realized that this was true.

But I've also learned that certain obstacles exist when attempting to apply Lean thinking to construction practices. Obviously, the traditional assembly line model doesn't translate cleanly to the construction world because many people in construction are staunch individualists—they don't like to think of themselves as assembly line workers; therefore, their eyes tend to glaze over at the very mention of concepts derived from an automated world. A second obstacle is that each project built is usually best described as either "one off" or "custom made," suggesting that implementing the type of systematic, repeatable practices that Lean is noted for won't readily apply. The third obstacle is pure pragmatism: due to conventional contracting methods, most construction entities struggle between competing for the limited resources at hand (i.e., money) and looking at ways to work in a cooperative manner to conserve and share these resources. The final obstacle involves the constantly rotating cast of diverse characters that populate each project—each coming with their own set of dynamics and personalities. This constant flux amid the usual pressures to produce in a timely manner tends to make people focus on immediate deliverables and personal differences rather than focusing on a more standardized, collaborative, and systematic way of approaching the work.

Fortunately, my experience has taught me that these obstacles can be cleared—and cleared with relative ease—if one knows how to do it. This book will take these obstacles into account and suggest how to overcome whatever resistance to Lean thinking might exist.

This book will help you think, in fresh ways, about what you do every day. To give you a sense of how your thinking might change for the better, here are some people-oriented assessments that you might never have made before. As you read through them, consider how much time (if any) you've spent on these assessments in the past, and how thinking differently about *how* your team works together will improve your overall efficiency:

Attitude: Does everyone on your team (particularly the leaders) care about succeeding? Are the leaders (and everyone else) ethical and willing to make tough and decisive calls? Do they care about the people they work with and are they committed to their success? Are they focused on eliminating waste and continuous improvement?

Planning: Is there an overall design, estimate, budget, and logic for the sequence of work and an organizational structure in place that people can orient to? Are execution and overarching philosophies in place?

Preparation: Does everyone understand the contract, the delivery system, scopes of work, roles and responsibilities (theirs and each others)? Does each person have a specific work plan?

Requisite skills: Do people know how to process RFIs, PCOs and submittals, and do they know how to log them? Can they track budgets and generate cost reports? If not, is there a plan in place to train them so they don't negatively impact the overall flow of the job?

Timing: Do people understand the schedule and have a working understanding of job priorities (keeping design and engineering out in front of construction), the importance of identifying long lead items, getting people paid on time, etc.? Are people working off of the current documents?

Execution: Can people make the best deal, get materials on site, manage to a budget, and accomplish a work plan (i.e., complete assignments in accordance with a specific deadline)? Is the work in place progressing to plan? Most importantly, can people do all of these things under pressure?

Responsibility/Accountability: Are people accountable for delivering what they promised, in a timely fashion, via their work plan? If they can't complete an assignment to plan, do they communicate this to the rest of the team or do they keep this vital information to themselves?

Communication and coordination: Are people constantly seeking to knit their disparate functions together? (That is, do engineers, PMs, superintendents, and general foreman walk the field together and look at drawings and contracts together?) Do people communicate where they are at, what they need, and when they will need it by—constantly? Is anything impinging on their ability to do this (e.g., poor leadership practices, lack of trust, etc.)?

Waste: Are there lots of reworks, issues falling through the cracks, duplications of service, or people producing less than what they could?

Process for continual improvement: Do the leaders constantly assess *how* the team is functioning? Do they seek team and owner feedback about what is "working" and what isn't? Do they examine processes to identify breakdowns in team function? Do they use staff meetings to identify and solve problems? Do the leaders encourage innovation? Do they use lessons learned and best practices?

As foreign as some of these questions may seem, they become second nature once you adopt the Lean principles that will be put forward in this book. Construction is a tough business for tough managers, and in the past, you may not have given many of these questions a moment's thought. You may rue that the business has changed, that you don't want to have to deal with all this "people stuff" on top of all the other craziness. But in a business intrinsically loaded with people and personalities, ineffective management structures and poor communication, thinking in a Lean way can make the difference between a profitable, competitive construction team and mass frustration and lost profit potential. Because many of the activities that occur on any given site are distinctly nonvalue added (i.e. things that don't add the actual construction process—like the paperwork requirements to mitigate risk), it behooves managers to be as efficient on the people side as humanly possible so they can get the most out of activities that *are* value added (contribute to the actual building of the project).

And believe me, I understand the craziness. A while back, I took a construction management and law class at a local community college. At the end of the class, the professor looked at me quizzically and said, "Gary, you're not a construction person, so I'm just curious—what precisely did you get out of taking this class?" I thought for a moment and gave the only answer that came to mind. "To be honest, between the potential lawsuits, onerous contracts, and the inherent risks, the only thing I keep wondering is why anyone in their right mind would want to do this for a living?"

So why do you do it? I know the answer many of you have given me, and it demonstrates that you're willing to tackle the people issues if the end result is higher productivity and less waste. There is something very pure about construction—something very tangible. When it's done right, you know it—and when it's done wrong, you know that too. But there is also something else. I've seen that wistful look in your eyes when a project finally emerges from being a pipe-and-conduit-filled hole in the ground— that point when the mythical and seemingly unattainable punch list starts to seem like less of a fantasy. There is something truly magical about being able to take a set of two-dimensional drawings and transform them into a living sculpture and know that you were a part of it.

But the deep satisfaction you derive from the technical side of construction isn't what you usually get from the people side of the business. Unlike rebar and concrete, people actually expect to be listened to and cared about. And if they make a mistake, they don't accept being torn down, ripped out, or replaced on the spot. I remember a fuming operations manager once saying, "You know, this would be a great business if it weren't for the people." He had just come from a job site where several of the staff had threatened to quit because they were frustrated with their boss. What they wanted from him seemed very simple. They were tired of toiling away without any clear direction or overall vision for the project. They wanted instruction on how to carry out certain procedures and wanted to understand how these procedures fit into the overall plan. And they wanted to know what was expected of them beyond what was listed in the procedures manual. Most importantly, they wanted to be engaged by their manager *before* he felt compelled to scream at them for something they had done wrong. The fix, in the operations manager's (OM's) mind, was so obvious: set a course, provide feedback, make yourself available for questions, and then stay engaged. A no brainer, right? So why did so many of his managers struggle when attempting to bridge the gap between the technical world and the world of people? Well, because in many ways, bridging this gap isn't that simple. The world of things is easy; you can beat them, cut them, pound them, shape them, and manipulate them to do just about anything that you want them to do. The world of people is much more complicated and convoluted. Or as Anthony Bourdain wrote in his best seller, *Kitchen Confidential*:

> Though I've spent half of my life watching people, guiding them, trying to anticipate their moods, motivations and actions, running from them, manipulating and being manipulated by them, they remain a mystery to me. People confuse me. Food doesn't. (2000, 299)

I believe that before you're finished with this book, you'll recognize that the bridge between the technical and the people side not only must be built, but also can be built. More than that, you'll see how the complex world of people can be managed effectively, even with its myriad personalities, time-specific demands, and uncompromising technical rigor. If you learn to think in Lean ways about how you approach your leadership, the people side will not only become less of a struggle, but also the culture that emerges will be the source of your future success—even if your project isn't contracted with integrated project delivery in mind.

One final note: Though their name is synonymous with mud right now, I will be referring to Lean practices originated by Toyota, simply because it is important that we not throw the baby out with the bathwater. Toyota's current quality problems are a direct result of their deviation *away* from the very Lean practices that helped them build the most reliable vehicles in the world. This temporary insanity that caused them to value production and market share over quality should not distract us from the fact that they became one of the greatest manufacturing enterprises in history because of the Lean methodologies that they pioneered.

About the Author

Gary Santorella, owner and president of Interactive Consulting, is a team building, strategic planning, and partnering facilitator, executive coach, and corporate instructor who has worked exclusively in the construction industry since 1996. He has pioneered a method of assessing construction teams at the job site and top organizational levels—in accordance to Lean principles—with the sole purpose of building efficient, flexible, and functional work groups. He has worked at sites across the country and in Europe. He has a BA in behavioral psychology from Providence College (1980), a master's degree in occupational social welfare from UC Berkeley (1990), and is also a licensed cognitive behavioral therapist, specializing in the treatment of PTSD and anxiety disorders, in the state of California. His client list includes Turner Construction, Rosendin Electric, Swinerton Builders, Ballast Nedam, Hensel Phelps, McCarthy Construction, Pan-Pacific Plumbing, the State Seismic Program of California (DGS), Vanir Construction, KCS West, Kajima Construction, St. Joseph's Health Care System of California, and the Port of Seattle.

For contact information, please go to his Web site at www.interactive-consulting.biz.

1

Lean Cuisine and Construction: The Benefits of a Food Industry Perspective

As you probably gleaned from the quotation in the introduction, I'm a huge Anthony Bourdain fan. But unlike *Kitchen Confidential*, this book isn't about exposing the underbelly of an industry. I referenced Bourdain because he is one of the best Lean thinkers on the planet (though he is probably wholly unaware of this). He learned about Lean thinking the hard way. By conducting a retrospective analysis of why so many of the restaurants he worked in failed, and incorporating the lessons that he learned from an imposing character he refers to as Bigfoot, he came to understand what was truly required to run a successful kitchen.

So between readings of *The Toyota Way* and various articles about Lean construction, and rereading *Kitchen Confidential*, it suddenly dawned on me why Lean methods sometimes failed to gain traction among construction professionals—even though these methods have been shown to be statistically sound. The world of construction and that of assembly lines seem diametrically opposed, so construction people (particularly managers) view themselves as operating in a very different environment from their counterparts in the manufacturing world.

On the other hand, construction professionals, head chefs, and managers of high-end restaurants have a lot in common. They both must deal with an amazingly similar array of issues on the people side of their work. In addition, construction projects and restaurants often fail or succeed for the same reasons—reasons that are rooted in Lean principles.

Consider some of the similarities between the two fields and how they are very different from a manufacturing operation.

Like most people who run kitchens, construction people are fiercely independent, and like to think of themselves as creative problem solvers. Both relish the challenge of pulling off the impossible every day and using their heads for something other than a hat rack. Construction managers and chefs alike enjoy being in control of their product every step of the way.

In contrast, we tend to think of assembly lines as fairly passive, system-driven places, where the opportunity for creative thinking and problem solving is limited. And, with assembly lines, at least on the surface, the process is what appears to be in control—not the people.

Like restaurants, job sites are loaded with colorful characters who are passionate about what they do, and this passion is often expressed in their interactions with others. To say the least, job sites and kitchens are highly dynamic places. Assembly lines? Not so much—at least not in the stereotypic sense. (Later on, you'll see how Lean-driven assembly lines differ from our stereotypic view of them and how dynamic they can actually be.)

Like their restaurant counterparts, construction managers have frequent contact with owners, end users (customers), workers, and suppliers. Assembly line managers tend to be a bit further removed from board members, stockholders, customers, and suppliers—and often have to work hard to gain access to these key players for their critical input.

Both restaurant and construction managers tend to be overly enamored with their subordinates' technical abilities, and tend to overlook their potential "team killing" behaviors in deference to the highly specialized products they are capable of producing. They are also usually horrified at the prospect of cutting loose "bad actors" for fear that they will not be able to replace their technical expertise. Assembly plant managers are actually much less worried about dealing with such issues. Team killers are usually dealt with swiftly and harshly. Because quality is built into the assembly line process, bad actors are exposed early on. And because much of the process is automated, and so many people are cross-trained, people on the line can be replaced without many major hiccups. Not so with kitchens and construction sites—at least not in the short run. And because most kitchens and job sites cut pretty close to the bone in terms of manpower, the loss of one key player is acutely felt.

Unlike assembly lines, which are often fully or semiautomated, both the restaurant and construction industry are fully dependent on *people* to produce a high-quality product. In both of these environments the people, in essence, *are* the product—and what constitutes subassemblies and finds its way into the production line is directly tied to what a real live person has generated.

Forgive me for belaboring these comparisons and contrasts, but I want to impress upon you what they communicate about Lean principles. When you compare kitchens and job sites and see the similarities, you recognize that success in both fields is dependent on the same qualities: preplanning, preparation, well-timed actions, constant communication, consistent execution, and so on. And when you get past the surface differences that these two fields have with a manufacturing plant, you realize that the plant also requires the same qualities to be in place to be successful. These qualities, of course, are the Lean principles. Take a look at Table 1.1, which describes the requirements for success in the food service and construction industries.

We often assume that a restaurant's success is due to its highly skillful and creative head chef. But great recipes and great ingredients, put together by skillful chefs, is only a small part of the equation. The truth is, when restaurants fail, in most cases, a creative head chef was at the helm who also utilized the freshest of ingredients. As Bourdain points out, the true secret behind the success of any dinner service doesn't reside solely with the creativity of the head chef—it rests on the kitchen staff's ability to create the same dishes consistently and efficiently, night after night, week after week. Like an assembly line, it is about being able to effectively carry out the same processes and procedures repeatedly. Ultimately, success in any kitchen is heavily predicated upon the kitchen team's ability to consistently communicate, coordinate their actions, and execute on deliverables—as a team. Failing to do so results in measurable waste in the form of some very expensive ingredients tossed into the dumpster. So, here is a key takeaway from the food world: it is not the ingredients alone or the technical prowess of the chefs that determine the excellent of a meal, but how effectively these ingredients are put together—*by all the people putting them together.* And all of these behaviors are, in turn, dependent on the planning and preparation that is required *before the meal is even cooked!*

And the same is true in construction.

Though the specifics (like the daily specials) may change, your team will need to fully understand their contracts, plans, and specifications (the menu). They will need to be able to vet requests for information (RFIs,) potential change orders (PCOs), and submittals (in accordance to high-quality standards), consistently log their actions, generate accurate budget reports, issue invoices, and make payments in accordance to an overall schedule. And if they are to be truly successful, just like in a restaurant, they can't perform these tasks in a vacuum.

TABLE 1.1

Key Elements for Success

Success Factor	Food World	Construction World
Attitude	In the business because they know the business. Willing to put tools in place for people to be successful. Create trust by having the backs of people who work hard. Have regard for people and company policies. Leaders expect everyone, including themselves, to live up to their commitments.	Leaders understand the big picture. Want to succeed and want the people under them to succeed. Willing to provide people with the tools to do the job. Create trust by having people's backs who produce. Have regard for people and company policies. Leaders expect everyone, including themselves, to live up to their commitments.
Planning	Menu and recipes in place. Overall design (French, Italian, Vietnamese etc.) in place. Estimate of ingredients needed (and budget) in place. Organizational structure and roles and responsibilities are clear.	Overall design in place. Contract in place. Estimate, budget, and logical sequence of work established. Plan for buy-out in place. Identification of long lead items. Organizational structure and roles and responsibilities in place.
Preparation	Everyone knows the menu cold. Frequently used ingredients are prepped in advance for each station. All needed ingredients have been preordered and are in the refrigerator. Everyone knows what everyone else is doing and is responsible for.	People know their scopes, plans, and specs, understand the contract delivery system, and have a work plan in place. Needed materials are preordered and are on site. Everyone knows what everyone else is doing and is responsible for.
Requisite skills	Know recipes and cooking techniques. Can chop, cut, dice, and sauté. Know proper meat temperatures to produce high-quality meals. Know how to track and monitor budgets and inventory.	Can process RFIs, PCOs, and submittals and log them. Can track costs and produce budget reports. Can manage to budget. Plan in place for training if skills are lacking. Know how to track and monitor budgets, materials, and manpower.

TABLE 1.1

(Continued)

Success Factor	Food World	Construction World
Timing	Appetizers precede entrees—desserts follow. Sequence of meat, vegetables, and sauces determined. Each station knows cooking durations and delivery time. Pay suppliers on time. Are working off of current menu. Actions are coordinated.	Know the schedule. Know to keep design and engineering ahead of construction. Mitigate impacts of long lead items and understand importance of getting people paid on time. Are working off of current documents. Actions are coordinated.
Execution	One thing to know what to do, another to actually do it—and do so under pressure. Best ingredients are purchased for the best price. Workstation managed properly. Everyone is accountable to deliver on commitments. Failed deadlines are not an option.	One thing to know what to do, another to actually do it. Best materials are purchased for the best price. Work plan managed properly. People are able to deliver under pressure. Everyone is accountable to deliver on commitments. Failed deadlines are not an option.
Communication and coordination	Constant communication regarding where they are, what is needed, by whom, and when. The goal is to knit together each workstation to produce a seamless dinner service. People ask for help when needed. Let others know if they are falling behind or have made a mistake.	Constant communication regarding where they are, what is needed, by whom, and when. The goal is to knit together the office and field functions to produce a seamless product. People ask for help when needed. People take responsibility to let others know if they are falling behind or have made a mistake.

(continued)

TABLE 1.1

(Continued)

Success Factor	Food World	Construction World
Waste	Issues identified before food is wasted. Goal is to eliminate replating. Effort is made to keep inventory at a minimum to reduce spoilage. Roles and expectations reclarified as needed. Eliminate deadwood and team killers. Productivity rate (successfully plated meals/waste ratio) is high.	Issues identified before they fall through the cracks. Goal is to eliminate redos or unwanted duplication of service. Effort is made to keep materials from the site until needed to prevent damage, reorders, and restocking. Roles and expectations reclarified as needed. Eliminate deadwood and team killers. Return on staff ratio is high.
Process for continuous improvement	Managers and chefs assess how the team is doing. Seek feedback from owner and customers. Seek feedback from employees in terms of frustrations or what can be done better. Probe why problems are occurring to mitigate root causes. Measure performance (number of successfully plated meals/waste ratio).	Managers assess how the team is doing. Seek feedback from owner and end users. Seek feedback from employees in terms of frustrations and on what can be done better. Probe why problems are occurring to mitigate root causes. Measure performance; increase return on staff ratio; reliability of commitments made via work plan (percent complete/deadline).

Communication and coordination between the office and field functions has to be frequent and constant. And similar to both kitchens and well-run assembly lines, the process needs to be predictable, repeatable, accurate, and consistently maintained. Otherwise, people-generated work will accumulate and, just like a plate of poorly timed food, will end up being tossed out as waste.

From a Lean perspective, it is also crucial to understand where kitchens and construction differ greatly from assembly lines: at the failure points. This is particularly true on the people side of both worlds. Assembly lines are fairly resistant to the vagaries of people. Unless someone truly screws up or sabotages the process, they run themselves and the product that comes out the other end remains relatively consistent in terms of quality. Not so with restaurants or job sites. A quick examination of Table 1.2 will demonstrate exactly what I mean.

TABLE 1.2

(Continued)

Failure Point	Food World	Construction World
Wrong attitude	Get into the business because friends told them that they gave great parties or because they want to show off their collection of antiques. No idea of business fundamentals. Spend wildly on the wrong things. Little regard for the people that work for them. Underreact or overreact when people fail to live up to commitments. Little focus on profitability.	Leaders have no understanding of big picture or particulars of the contract. Track technical issues only—spend no time focusing on *how* to get to the result as a team—because they are not sure about the steps required to achieve the result they are seeking. Little regard for others. Underreact or overreact when people fail to live up to commitments. Little focus on profitability in favor of "just getting it built."
Absent planning	No consistency. Menu changes constantly. Theme changes constantly (one day country French, next week southern Italian). Poor planning for procurement of ingredients (buy way too much or too little—and at a bad price). Budget is a mystery. Nobody clearly knows who is doing what.	Poor understanding of project documents, drawings and delivery system. The schedule changes constantly. No plan in place to track the budget. Work plans are a mystery. Long lead items missed. Clear organizational structure and roles and responsibilities are completely lacking. No one knows who is doing what.
Poor preparation	Wait staff don't know the menu. Prep work not done. Workstations are a mess. The right hand doesn't know what the left is doing. Key ingredients are missing or not purchased. People begin to quit because they don't like the way they are being treated and have no confidence in their ability to be successful.	People don't know their scopes, plans, and specs, or understand the contract delivery system. No work plans are in place. Trailer is a mess. Needed materials not brought out or not on site. Actions not happening in accordance to schedule. People begin to act out (become sullen, passive-aggressive, hostile) because they don't like the way they are being treated and have no confidence in their ability to be successful.

(continued)

TABLE 1.2

(Continued)

Failure Point	Food World	Construction World
Requisite skills lacking	Don't know how to chop, cut, dice, and sauté—or if they do know, they can't do it under pressure. Don't know how to track and monitor budgets and inventory. No plan in place for training if skills are lacking.	Don't know how to process RFIs, PCOs, and submittals and log them (at least in terms of their current company's standards). Don't know how to track costs and produce budget reports. Manage budgets haphazardly. No plan in place for training if skills are lacking.
Poor sense of timing	Appetizers come out at the same time as entrees. Sequence and timing of meat, vegetables, and sauces is a mess. Each station only focuses on its own needs, could care less about what is happening on other stations. No sense of priorities (what is important right now). Suppliers paid late or not at all. Communication is thoroughly lacking.	Don't know the schedule. Construction gets ahead of engineering or design. People focus on their own areas of expertise and show little regard for others. No regard for teammates' needs in terms of timing. No sense of priorities (what is important right now). Miss long lead items. No coordination. Subs running up each other's backs. Not working off of current documents. Subs and vendors not getting paid. Communication thoroughly lacking.
Sloppy execution	Start to cut corners to save money. Lots of yelling and screaming—or simply quitting without notice. Can't produce under pressure. Workstations mismanaged. No leadership direction. Commitments drop and deadlines fail as people become more and more frustrated with each other and focus only on themselves and their needs.	Start to increase overtime to maintain schedule. People do not know what is expected of them or how to coordinate their actions. Leadership direction is nonexistent. Work plan dropped and tasks not completed. Commitments drop and deadlines fail as people become more and more frustrated with each other and focus only on themselves and their own needs.

TABLE 1.2

(Continued)

Failure Point	Food World	Construction World
Lack of communication and coordination	Communication nonexistent. Lots of blaming and withholding as people scramble to make themselves look good. Meals not plated due to lack of coordination. No one asks for help (or gives help) if needed. No one admits to making mistakes.	Communication nonexistent. Lots of blaming and withholding of information as people scramble to make themselves look good. No coordination between field and engineering. No one asks for help (or gives help) if needed. No one admits to making mistakes.
Waste	Tons of wasted food. Inventory is not tracked. Lots of spoilage. Lots of replating because of poor coordination. Who is doing what is never clarified. Team-killing behaviors are allowed to flourish. Restaurant bleeds money like a sieve.	Tons of waste in terms of redos or unwanted duplication of service. Issues fall through the cracks due to poor communication, lack of coordination, and not following through on commitments. Team-killing behaviors (blaming or withholding information/ dropping deadlines) are allowed to flourish. Project bleeds money like a sieve.
No process for continuous improvement	No attention paid to customer complaints. Change menu or theme without ferreting out root causes for previous failures. Don't ask employees how to improve things. High turnover. Restaurant becomes one of the six out of ten that fails. Number of successfully plated meals/waste ratio is low.	No attention paid to owner or employee complaints. People continue to work in isolated silos. No probing as to why problems are occurring to get to root causes. GC or subcontractors are thrown off the job. People quit in droves. Project ends up in litigation. Return on staff ratio is extremely low.

As you can clearly see, the issues that kill kitchens and job sites alike are uncannily similar—and are almost entirely people driven. It is about the failure to plan, prepare, communicate, execute, and deliver on what was promised. This is why being a Lean thinker and focusing on creating a Lean culture is so critical. Anything that is inadequately planned, that interrupts the smooth flow of work, or is an obstacle for the people carrying out the work, can and will equate to commensurate failures in quality, productivity, and profitability. Because construction is populated by hard-driving, results-driven—go, go, go—types of individuals, the errors that occur due to people issues will have a multiplicative effect in terms of waste on the technical side. In the "blow and go" environment of construction, if poor communication, ineffective timing, inadequate execution or training, and ineffective conflict resolution are allowed to flourish, the waste generated by such problems (and the cost to reverse them) will be compounded exponentially in the form of lost profits.

To mitigate these impacts you could try to "engineer" your way out of these people-generated problems, which is precisely why Lean construction practices such as building information modeling (BIM), prefabrication, pull planning, last planner systems, and reverse phase schedules are gaining in popularity. But these process improvements will only serve to alleviate part of the problem. When it comes right down to it, success on any job site will always reside on the leader's ability to handle the people side effectively. This means applying Lean principles to people, and that's exactly what the following chapters will teach you how to do.

2

The Lean Team Challenge

A few years ago, my wife and I, in what only can be described as a temporary bout of insanity, embarked on a two-week driving tour of England. After adjusting to the little mirror box of horrors (steering wheel on the right, gear shifter and rearview mirror on the left), better known as a Ford Mondeo, we left the relative safety of London and headed for the English countryside. After a few hours of white-knuckled driving we reached our destination—the mother of all Western European building projects—Stonehenge.

What strikes you immediately about Stonehenge is the phenomenal scale of the site. Though the actual stonework structure is a modest 320 feet in diameter, the entire site spans hundreds of acres. Besides the well-known circular stone structure, the site is comprised of a complex series of earth-works, burial mounds, and avenues, as well as remnants of an intricate lattice of scaffolding and connecting wooden buildings. But while most people who study the site ponder its significance (was it a spiritual hub? an elaborate burial site? a prescientific means to predict moon phases and the seasons more accurately?), I couldn't help marveling at the fact that it had been built at all. It is estimated that some 20 million man-hours went into the construction of Stonehenge!

So, who were the construction leaders on this site anyway? And what precisely were they able to convey with such conviction they were able to convince their fellow villagers to leave the relative comfort of their wooden huts, and over the course of seventy-five generations (5000 BC to 2600 BC), drag sixty or so 25-ton sarsen and bluestone slabs 19 miles across the Salisbury plain and erect them to precise specifications? Why is it, that after a couple of centuries, people didn't just say to heck with this! That's what I really want to know. Don't you?

While we are social beings, we are also overwhelmingly hardwired for self-interest. Huge portions of our brain are specifically devoted to autonomic stress responses that compel us to abandon the greater good in favor of self-preservation. The struggle for existence, as Darwin described it, is just that. Whether at work or driving in our cars on the way to work, we are constantly looking for opportunities to enhance our individual position or scanning the horizon for potential danger. This is the reality you and I know all too well and what every team leader struggles against on a daily basis.

So by what means did these ancient leaders overcome these natural tendencies? What did they say? What did they do? Unfortunately, we'll never know. Unlike the structures they left behind, the motivational tools that they used to get the job completed are lost forever. But perhaps we can draw some inferences from current job sites.

Today, when a construction project doesn't go well, most higher-ups point the finger at the leader. The inference is clear: the underlying belief is that people are naturally inclined to *want* to work together, and if they don't, it must be due to the incompetence, inattention, or interpersonal shortcomings of the leader. But this is not exactly how the human equation works. Even when teams are firing on all cylinders, most people don't cease being individuals within the team environment. They are constantly scanning to see whether or not it is in their best interest to continue to throw in their lot with their teammates in the form of taking on extra work or eschewing individual recognition, or whether it is better to separate themselves from the pack, presumably where individual recognition for their efforts can be maximized and reduction of negative impacts from others can be minimized.

It is far more accurate to say that when things aren't going well, what the leader has failed to do is to give his or her staff a compelling reason to drop their biologically determined default position of "me" centeredness, to one where they perceive a value in throwing in their lot to accomplish some collective greater good.

Here is a key point that leaders need to grasp, but seldom do: if we are going to ask people to willingly give their best and contribute fully to team and project goals—and endure the extra work, long hours, fatigue, and increased anxiety that come along with it—there has to be something in it for them besides a paycheck.

With enough muscle, anybody can make someone do something against his or her will. All that is required is the ever-present dread of punishment. Stalin, Hitler, Mao, et al. were masters at this. (Though admittedly, they were also skillful at convincing others to engage in state-sanctioned atrocities for the attainment of individual benefits via the creation of an idealized society.) Clearly, people can be forced to do just about anything if they believe that their survival is at stake. But getting people to *want* to perform collectively toward some desired end requires a leader to tap into a very different understanding of the human condition. When they are able to do so, most leaders arrive at a paradoxical epiphany. Instead of carrying a big stick, they arm themselves with the most powerful leadership tool of all—an invitation!

Rather than trying to work against their instincts, or relying on endless threats, effective leaders invite their staff to *want* to give their best by understanding what it is that people want and need in exchange. Employees want

- A sense of purpose—they need to know that what they are being asked to do has significance—that they aren't just cogs in a meaningless, incomprehensible wheel
- A sense of accomplishment—they want to feel that what they are doing makes an actual contribution toward a greater good
- A sense of intellectual or personal skill development—that they are learning a new skill or way of thinking that will serve them in the future
- A meaningful voice—that their informed opinions will be sought out and listened to
- Recognition—that they are respected and held in high esteem for making important contributions and sacrifices for the greater good
- A sense of equity—that everyone on the team will be held to the same standards
- Financial benefit—that what they are doing will lead to financial gain now or in the future
- Responsibility—that when earned, they will be trusted to carry out and accomplish a variety of important functions within the organization with minimal supervision and will be recognized and rewarded for doing so
- The belief that if they are struggling, or have worries and concerns, they will be given meaningful support to help them succeed
- That the problems they encounter will be viewed as team problems rather than individual failings, and that they can expect support and cooperation from their teammates and superiors should they struggle—without fear of belittlement or retribution

What the leader is saying via such an invitation is this: "If you are willing to work hard, share information, execute your duties, help your teammates be successful, and drop your self-interested stance, I will make sure
that monetarily, professionally, intellectually, and psychologically you
are recognized and rewarded for doing so." In addition, the leader gives
unspoken assurances of equity, that is, that high standards will be maintained, and that those who maintain high standards will be rewarded and
those who violate these standards will either be punished or, at the very
least, not rewarded. Not surprisingly, these are also the key elements for
ensuring uninterrupted workflow in a Lean culture. Why do I say this?
Because when people feel that the leader is already doing his or her best to
provide them with what they need, they no longer feel the internal pull to
fight for things such as recognition or status. When their needs are being
met, they will reduce their vigilance, drop their self-protective stance, and
will be better able to focus on desired results. But if they feel that the leader
is not invested in fulfilling their needs or interests, they will revert to self-
interest, pull away from the team, fail to coordinate their actions or share
information in favor of making themselves look good, and the predictable
disruptions to flow and execution will result.

Given the face validity of the above assertions, you may well ask why so
many leaders fail to attend to the needs of their staff.

Usually, such failures are the result of factors that are all too common
in construction: information overload and stress. Let's look at an example. Let's say that a project manager, due to being named late to a project,
barricades himself in his office in order to review project documents and
only emerges when called upon to attend a plethora of meetings with the
owner. What happens if this same leader, so stressed by what he discov-
ers in the documents, or so overwhelmed by the deliverables generated
in the owner meetings, either avoids engaging with the team or simply
fails to acknowledge the contributions his staff are making to the overall
team effort? Again, because people are constantly scanning whether it is
better to remain a part of the team or to look out for themselves, if the
leader intentionally or unintentionally withholds or reneges on the promises implied in the invitation, or otherwise fails to attend to staff needs,
people on the team will revert to their self-interested positions. Worse,
because they will often feel betrayed, duped, or foolish for abandoning
their self-protective stance in the first place, they will actively resist the
efforts and decisions of the leader in the future—even if these efforts are

objectively viewed as well intended. Some leaders think that a clever way to avoid this type of potential failure is to simply extend no invitations at all. Sorry to say, but once you accept the mantle of leader, these promises, and the expectations contained within, are already implied. If you choose to ignore this reality you do so at your own peril.

Now do you see why so many powerful leaders throughout history have resorted to the business end of a gun? These aren't easy promises to live up to! But the point of the above example is to demonstrate just how easy it is to derail the smooth flow of work by simply ignoring the important relationship that leaders have with their staff. You'll notice in the example given that there was no ill intent on the leader's part to negatively impact his team. In fact, the intent was probably the exact opposite—to gain control of project particulars in order to assist his staff down the road. But, in terms of actual impact, it boils down to this: people have needs, and they expect to have these needs met through the actions of their leader. If they don't, they will attempt to get their needs met through some other means. You can readily see how fragile this whole notion of team truly is.

Even though we throw around the term *teamwork* all the time, we often fail to grasp the leadership implications. If a leader simply has a bad day or in the heat of the moment inappropriately loses his or her temper, and does nothing to repair the resulting damage, months of hard work to establish a functional team environment can be destroyed.

It is essential for leaders to understand that the invitation espoused at the beginning of a project is not one that is easily said and then quickly forgotten. Such lip service does more damage than if nothing had been said at all. Truly effective Lean leaders quickly realize that everything done and said each day is an invitation for individuals to stay invested and committed to the overall team process throughout the life of the project.

As daunting as this may sound, in truth, it is the exact opposite. Each day is an opportunity to invite your team to stay engaged. Lean leadership requires a full-time mindset, but only a part-time investment in behavior. Every question answered, every direction thoughtfully dispatched, every worry that is listened to and acknowledged is, in fact, a delivery on the promise—and will keep your team invested and engaged. Rather than looking for the exits, your people will look for new ways to make contributions to the overall team effort.

But one caveat: A functional Lean team process won't emerge overnight. It requires one more key element of human understanding—and a good dose of patience. There is an evolutionary progression that happens for most individuals within a team. When people first come to a job, they initially bond to a small group of people who they happen to like or with whom they share a common personal interest. As time goes on, and as they gain a sense of purpose and meaning through their work, they will bond to their immediate work group (i.e., field work, engineering, accounting, etc.). As they acquire greater knowledge and learn more about the big picture, they will come to see how what they do contributes to the overall project goals and will then begin to bond to the entire project team. And as they gain a greater sense of the overall business plan, and feel a connection to the overall scheme, they will begin to bond with an entire division or business unit. And at the very end, as the company's vision and mission become clearer, and as their personal job opportunities and responsibilities increase in accordance to this vision, they will then bond to the entire company.

Whether due to their own internal ambivalence or the failure of leaders to extend meaningful invitations, not everyone will climb this progressive ladder of belonging. But the more attention you pay to what is important to people, and the more you invite them to be engaged and involved in the team process, the more likely it will be that your subordinates will *want* to be an active part of what is happening. More importantly, they will *want* to do whatever is necessary to make contributions to the greater good of both the project and the company.

As a prelude to the next chapter, and to demonstrate just how simple yet vital this notion of extending invitations truly is, I'd like to introduce you to my own model, which I call the "brand your cattle and build your fences" model (Figure 2.1).

As an example, let's take the case of a contractor who is executing work on site, whose leadership team is comprised of a project manager (PM), a purchasing agent (PA), and a general foreman (GF). Intuitively, we know that for this team to be productive, and to reduce potentially wasteful outcomes, they will need to engage in healthy exchanges of communication. Not only will they need to understand their own job duties and be able to execute their own role, but each team member will also need to have a working knowledge of their teammate's tasks and responsibilities. Additionally, they will need to exchange critical pieces of information, develop a mechanism to identify potential problems, and establish a sense of timing for key deliverables so that their individual actions can be coordinated with maximum efficiency. In other words, while each person owns

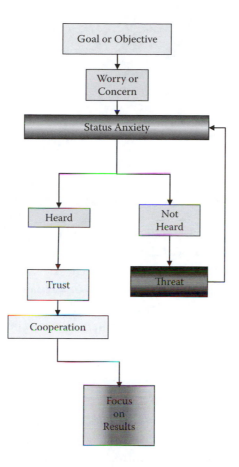

FIGURE 2.1
"Brand your cattle and build your fences" model.

his or her own role or circle of responsibility, critical interface points need to be established to ensure that the work product that each person contributes is successfully executed as an overall desired outcome. Graphically, it would look as shown in Figure 2.2.

The areas of overlap are where teamwork is required. And here is where the whole notion of invitations comes into play. To solidify these interface points, it is imperative that the leader extend an invitation for everyone to be open about the problems that they are encountering and to ask for help, so if issues do arise, they can be handled as a team.

So, what happens in this scenario if one of the people in this triad encounters a problem that impedes his or her work? Let's say a cost problem

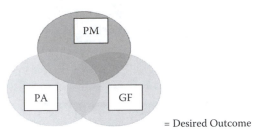

= Desired Outcome

FIGURE 2.2
Functional team.

involving manpower emerges that the GF can't solve on his own (e.g., the foreman that the GF oversees is bullheaded and doesn't buy into the man-loading schedule that the management team has developed and believes that any cost overruns caused by overstaffing can be made up in the buy-out). Let's also say that the GF has made an initial attempt to "invite" the fore-man to his way of thinking but the foreman still won't budge and continues to overman the job. It should be obvious that the longer this problem goes unfixed, the more money will drain out of the profits, and the more the GF will continue to worry about it. This is the state that I call status anxi-ety. Status anxiety is far beyond the run of the anxiety. It's that point when people start worrying about how their teammates are viewing them if they unable to get something to happen (i.e., "Will the PM think less of me if I can't get this foreman under control?" "Could I lose my job over it?")—hence the title "status anxiety." We become so fixated with how we look in the eyes of others that we lose sight of what we are trying to achieve.

Going back to our example, if the GF takes the risk to voice the problem—as initially invited—and the PM and purchasing manager not only hear him, but roll up their sleeves and develop a plan to address the issue as a team (i.e., decide to have a joint meeting with the foreman to thoroughly go over the plan, discover his resistance points or lack of understanding, seek out his input and buy-in, or, as a last resort, remove him from the project), then the GF will no longer remain in status anxiety mode. He will move away from his preoccupation with self (how he is being perceived) and, through a renewed sense of trust and cooperation with his teammates, will focus entirely on achieving the desired result.

But let's look at what happens if this situation were to be reversed. Let's suppose that the GF took a risk to admit that he is struggling with the foreman, and instead of receiving support, the PM said, "Hey, I've got my own problems to deal with. Getting the foreman under control is your job. If you're not up to the task, maybe I should find somebody who is."

Where will the GF's status anxiety go then? Will it reduce or increase? And as a result, where will his focus be? In the previous example, being heard allows the person who is struggling to move away from self-interest and instead focus on desired results. Feeling unheard produces the exact opposite result. Rather than feeling invited to share the load with his managing teammates, the GF will actually feel barred from the party. Through his actions, the PM will have *uninvited* the GF from the team process and, as a result, unwittingly extended an open invitation for him to suffer in silence. Now, here is a key question: In the midst of status anxiety, and feeling cut off from his managing partners, how will the GF look upon the foreman? Will the foreman still be viewed as a potential teammate by the GF, that is, someone to be engaged with and, in turn, invited into the team process? No! The foreman's resistance will no longer seem like a mere difference of opinion; it will be perceived as a threat to the GF's existence—something to be attacked, micromanaged, or actively avoided. Rather than trying to elicit buy-in, the GF will look upon the foreman as an enemy; he will become ever more preoccupied with his own status anxiety and will likely go on the attack. Worse, he will look unfavorably at his managing partners, also perceiving them as threats rather than as potential allies. It is precisely at this point that most people in this type of situation make a critical decision that is lethal to the Lean team process: they decide to pull their circle of responsibility away from their teammates, and instead of interfacing, exchanging ideas, and coordinating their actions, go it alone. They focus on whatever makes them look good regardless of the impact this stance may have on the overall project outcome. From this self-protective posture, a number of Lean-killing and CYA behaviors are spawned. Accusatory emails are hurled, information is hoarded, and mistakes are denied, all in an attempt to protect one's own position—hence the model's name. Instead of working together, people under threat build their fences, brand their cattle, and fiercely defend their own turf while only peripherally attending to a desired project outcome—even worse, they will feel completely justified for engaging in these Lean-killing behaviors (Figure 2.3)!

Here is the critical question to ask as a Lean thinking leader attempting to build a Lean culture: What is the cost in real dollars in terms of productivity

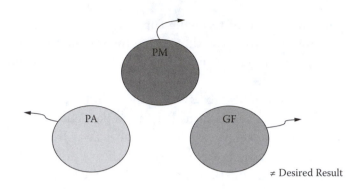

≠ Desired Result

FIGURE 2.3
Branded team.

and waste when people decide to pull their circles of responsibility away from the team and go it alone? In today's highly competitive environment, where most jobs are estimated extremely close to the bone, any issue that isn't coordinated with maximum efficiency, any work that has to be redone as a result of a failure to share critical information, any problem or concern that is held on to because a rebuke has impelled someone to remove himself or herself from the team process can and will turn a potentially profitable job into a write-down in a New York minute! There is a direct cost whenever a trailer full of people performs as individuals rather than a cohesive team. I can't emphasize this point strongly enough: $1 lost because people were unable to manage their professional relationships is $1 too much!

There are also indirect costs to factor in as well. Harkening back to our profit equation (Profit = Price – Cost × Volume), there is also a loss of profits associated with decreased volume. In the above example, after witnessing the subcontractor's inability to execute an effective manpower plan as a team, how likely is it that the owner or general contractor will approve a submitted change order or claim—even if the claim or change order is valid? Rightly or wrongly, the assumption will be that the issue became a claim or change specifically because of poor teamwork on the part of the subcontractor—rather than due to a change in contract documents.

In addition, what is the likelihood that the same subcontractor, if all bids are equal, will gain future work with the owner and GC if they are perceived as being unable to manage their own team process with maximum efficiency?

Think about it: Due to something as simple as a rescinded invitation, both current and future profits can and will be lost!

If you are truly going to think of yourself as a Lean leader, you will need to adopt the mindset that people issues are never just simple asides to the real work of construction; they are integrally connected to the work—in real dollars—now and in the future. It is vital that you see yourself not only as the possessor of a set of critical technical skills, but also as the key extender and holder of an invitation. Because it is through your delivery of this invitation that effective team process happens. If you take only one thing away from this chapter, I hope it is this: it is your responsibility to make sure that your people are always pushing their circles of responsibility together—and that they truly understand that there is something in it for them for doing so. They can disagree, they can argue—they can even get passionate about it—but at the end of the day, no one on the team has permission to go it alone, because if they do, everyone loses. In this context, stop thinking of yourself as a manager of individuals who are performing discrete tasks. Instead, think of yourself as a manager of interactions *between individuals*. It is your job to ensure that the necessary interfaces that need to occur actually do occur. Lean culture happens at the intersection *between* individual responsibilities. And the best way to make sure this occurs is by living up to the promises implied in your leadership invitation!

THE INVITATION TEST

Here's a quick and easy way to tell if you are doing what it takes to invite people to be a part of the team. During your next staff meeting, ask someone you trust to evaluate you on how well you did the following (have them circle yes or no):

1.	Did you use the word *we* instead of *I*?	Yes	No
2.	Did you truly encourage people to ask questions?	Yes	No
3.	Did you answer your teammate's questions?	Yes	No
4.	Did you refrain from cutting someone off so you could continue talking?	Yes	No
5.	Did you clearly identify upcoming milestones?	Yes	No

6.	Did you show more interest in getting to the milestone as a team versus simply driving to the result?	Yes	No
7	Did you encourage people to speak up about their worries or concerns?	Yes	No
8.	Did you respond empathetically to people's worries and concerns about reaching the milestone?	Yes	No
9.	Did you ask people what they would need from you or their teammates to help them to accomplish the milestone?	Yes	No
10.	Did you ask for people's input or help to solve an issue, concern, or possible roadblock?	Yes	No
11.	In an hour-long staff meeting, did you refrain from speaking for more than twenty minutes?	Yes	No

If the answer was no on more than three of these items, you've got some work to do on your invitation skills.

3

Trust—Laying the Foundation

After you have extended your invitation, and fully grasped its implications, you may be wondering, what comes next? How do I sustain high levels of productivity and teamwork even during difficult times? The key can be summed up in one word: trust.

Trust is another one of those words that we throw around a lot, assuming that its meaning is clear. But what does *trust* truly mean, particularly on a job site?

In his seminal work, *The Five Dysfunctions of a Team*, Patrick Lencioni states that there are two primary components required for a team to consider itself trusting: benefit of the doubt and vulnerability. He says that it is crucial, particularly in times of trouble, that each person on the team give each other the benefit of the doubt and not doubt one another's heart or motivation. If someone on the team should fail to execute correctly, his or her teammates need to assume that his or her intentions were good (i.e., "Gary made a prioritization error and failed to update the submittal log on time" versus "Gary doesn't give a damn"). On teams where trust is high, the focus is on finding solutions to problems—as a team—not ferreting out scapegoats.

The second key component is a sense of vulnerability. Simply put, this means that everyone on the team is willing to risk personal safety for the sake of honesty. If they make a mistake, they own it; if they need help, they ask for it; if they don't understand something, they say so; if they have an interpersonal or technical shortcoming, they admit it. Rather than covering their posteriors or engaging in a number of behaviors to create the impression that things are going well, each person on the team takes responsibility and stays focused on reality—even if that current reality reflects badly upon them. Similar to the power behind team invitations, the reason they do so is that they see a value in doing so. When leaders encourage and honor vulnerability, their people will come to believe that the best way to be a valued team member and to ensure uninterrupted workflow is to be completely forthcoming.

Given that this is counterintuitive to their hardwiring, the important question to ask at this point is *why* they have come to believe this.

Behind closed doors, I've listened to leaders spend a great deal of time speculating as to whether or not they can trust the people under them. But what surprises them is when I turn their question around and ask, "How do you think your people gauge whether or not they can trust you?"

The answer to this question, as well as to why people come to believe that honesty is indeed the best policy, is largely based on how the leader handles situations when someone on the team has let the side down. What people assess is this: Do the leaders tend to treat every mistake as if it were the end of the world? Do they feel the need to rub a person's nose in his or her failures to make sure that he or she "gets it"? Do they react harshly whenever someone musters up the courage to deliver bad news? Do they commence to micromanage to such an extent that they crush all vestiges of independent thinking and initiative? If the answer to any of these questions is a resounding "yes," it's not much of a leap to conclude that vulnerability and trust on that team will be lacking. Rather than coming to believe, as Lencioni states, "that there is no reason to be careful or protective around the group," when leaders use people's vulnerabilities against them, this is experienced as psychological punishment, and people will quickly learn to associate their leaders with psychological pain. Soon after, they will learn that the best way to mitigate such pain is to bury mistakes, avoid all forms of responsibility, and do whatever it takes to get off the leader's radar screen. *Self-protection* becomes the watchword for such teams. Since the leaders are perceived as a threat to their well-being, they become preoccupied with building their fences, branding their cattle, and otherwise attempting to manage their manager's reactions, rather than focusing on what they need to do to improve. As a result, present and future productivity suffers. As Lencioni summarizes,

> Members of teams with an absence of trust…
> - Conceal their weaknesses and mistakes from one another
> - Hesitate to ask for help or provide constructive feedback
> - Hesitate to offer help outside of their own areas of responsibility
> - Jump to conclusions about the intentions and aptitudes of others without attempting to clarify them
> - Waste time and energy managing their behaviors for effect

- Hold grudges
- Dread meetings and find reasons to avoid spending time together

And, I will add, they will also feel completely justified in doing so.

Conversely, when leaders provide support, help them to learn from their mistakes (viewing mistakes as part of the learning process), thank and honor their staff for coming forward to voice problems and concerns, and are quick to point out past successes in order to bolster their confidence, people learn that vulnerability is not only valued, but also is the very basis for the team's success. More importantly, they will attend to what they need to do to improve so that the team's efforts can be executed with maximum efficiency—even if it means that, at times, they will expose themselves as being less than perfect. As an adjunct, people in an environment where vulnerability is encouraged actually take on more responsibility, rather than shrinking from it.

Referring again to Lencioni's summary.

> Members of trusting teams ...
>
> - Admit weaknesses and mistakes (interpersonal and technically based)
> - Ask for help
> - Accept questions and input about their areas of responsibility
> - Give one another the benefit of the doubt before arriving at a negative conclusion
> - Take risks in offering feedback and assistance
> - Appreciate and tap into one another's skills and experience
> - Focus time and energy on important issues, not politics
> - Offer and accept apologies without hesitation
> - Look forward to meetings and other opportunities to work as a group

So how does this relate back to the concept of invitations? It has been my experience that people make the determination of which type of invitee they are going to be—trusting or self-protecting—within the first few days of a job. The bad news is that once the notion of self-protectionism takes hold, over time, it becomes more and more resistant to change. Therefore, it is important for Lean leaders not only to invite people to be a part of a team, but also to invite them to be vulnerable. If continuous improvement is your aim, an environment for honest discussion and self-evaluation, with no holding back, must be created and maintained by the leaders.

Given the nature of this industry, it probably seems counterintuitive to encourage vulnerability. A brief examination of the average construction contract is replete with threats of punishment for failure to perform. Therefore, encouraging people to openly own up to their mistakes, on its face, might seem like a recommendation for suicide. I readily acknowledge that, at least externally, there will be those who will be quick to exploit any missteps that you or your staff admits to. But the important question to ask yourself is this: Do you really want to replicate this same feeling inside of your own trailer? Do you want people walking on eggshells or otherwise inculcating the belief that it is better to hide mistakes and shortcomings within their own trailer? Wouldn't you rather have people stepping up—honestly and directly? Again, this isn't just a touchy-feely issue—it's a dollars and cents issue. Whenever people choose to put time and energy into hiding problems or mistakes, or pretend to know something that they don't, what are they *not* spending time focusing on and what types of errors are they likely to make as a result? Wouldn't it be far better not to have to sift through all the smoke and mirrors and deal honestly and directly with the problems at hand? The reality is that a sense of invulnerability breeds waste in the form of misfires and lost productivity.

It is important at this point for you to honestly assess whether or not you are promoting the type of environment that allows vulnerability to occur. Let's take an example: Suppose you are a project manager for a general contractor, and your superintendent has just come to you and told you that her concrete subcontractor has informed her that they are having problems in their prefab yard, and form work will be delayed three days. You could respond by saying: "Shelley, I'm really not happy about this, but I'm glad you told me right away. Do we have a recovery plan so we can stay on schedule?" Or you could say: "Damn it, Shelley! How could you have let this happen?"

Can you feel the different impact each of these responses has on trust and vulnerability? The first question assumes good intentions, honors vulnerability, and invites active problem solving. The second, while perhaps feeling good to express in the moment, is a siren song for Shelley and everyone else on the team to run for the hills! Rather than inviting her trust, the latter statement invites defensiveness and avoidance of accountability.

There is another issue regarding trust that is rarely discussed. Some of you may feel that you do a pretty good job promoting vulnerability, yet you still hear whispers that trust on the team is lacking. There is something else that you need to consider, and it has to do with the unique engineering environment that you work in. Construction is a process-driven industry. There is a procedure for almost every job site activity. To be successful, your staff is highly dependent on procedural clarity in order to do their jobs effectively. Unfortunately, some construction leaders view the need for producing clear job descriptions, a functional organizational chart, a well-understood and respected chain of command, publicly posted and color-coded schedules, and relevant and well-maintained procedures manuals as trivial, mundane, or perfunctory tasks that get in the way of the real work. A word of warning: If such basics (you'll read much more about these later on) are not provided, their absence will negatively impact your team and, over time, undermine the team's trust in you. If you are preoccupied with more pressing tasks, your staff will give you the benefit of the doubt—for a little while. But as they flail about in a sea of uncertainty, and as their own mistakes begin to pile up, who do you think they will view as the source of their failure? All they will know is that when they needed a lifeline in the form of a codified procedural pathway, all they received from you was a void. It won't matter if you intended this to happen or not. Voids such as these are rarely filled in positively. They are not going to say, "Oh, poor Gary—I guess he was too busy attending meetings to put together a procedures manual." Instead, they will say, "That damn Gary doesn't give us the tools we need to do our jobs! What the heck does he do all day?" Now, you could protest at this point and countercharge that your staff is failing to give *you* the benefit of the doubt—and of course you'd be right. But that misses the essential point. Think about it; when you have been in their shoes, didn't you draw the same conclusion about your own boss when you were struggling? Instinctively, we link our success directly to the actions of the leader—and for good reason. From whom else do we get what we need to be successful but through those above us in the food chain? Having the sense that we can be successful at our jobs is vital to our sense of well-being; after all, in real terms, success determines whether or not we can buy a house, send our kids to good schools, pay our medical bills, enhance our careers, and in general, bring good things into our lives. So quite literally, any voids that create a perceived barrier to our success will be considered a threat. And by now, you understand all too well the implications of putting people in such a state. Threats, and the fear they generate, are the biological opposite to a sense of trust.

One last critical point to make about trust, and it is particularly salient in the high-pressured, fast-paced, high-stakes world of construction. We all blow it from time to time. Yelling, shutting down, withholding information, arriving late to meetings, failing to give people our undivided attention—the list of things that leaders can do under stress that can damage trust seems virtually endless. But the key question is this: Do you care enough to recognize when you have blown it, and are you willing to take responsibility for it? Your staff needs to know that you care enough about them to hold yourself accountable. They don't expect you to be perfect. But they do expect you to respond when you do something that could jeopardize vulnerability.

Remember, the best way to encourage vulnerability is to model it yourself! Most people, sadly, weren't upset that Bill Clinton had had an affair. What they were upset about was that when he had a chance to step up and take responsibility for it, he didn't. He will forever be remembered for uttering the words, "I did not have sex with that woman." Conversely, when JFK got up in front of the American people and took full responsibility for the Bay of Pigs fiasco, trust in his presidency actually went up!

4

Is Your Attitude an Advantage?

In my experience, the single most common root cause for the perpetuation of errors and the incursion of waste at the job site level is leaders who exhibit poor attitudes. In this chapter, we'll discuss issues concerning attitude that directly effect creating a Lean culture. Some of these, at first blush, might not seem so readily apparent. But an increased awareness of the subtle ways that your attitude can affect your team will help you to move forward with your Lean leadership efforts.

Our attitude is the well from which we draw our invitations of trust, and in turn is the source point for the promotion of a culture that focuses on continuous improvement. But first, what do we mean by *attitude*? It is a word that we toss about with seeming certainty, having both good and bad connotations, but what does it really mean? *Attitude* is cryptic shorthand for the beliefs that we hold about others in relation to ourselves, and the constellation of justifications we use when acting upon these beliefs. In short, what we believe about others in relation to ourselves governs how we act toward them; our attitudes are the projection of those feelings.

When we are leading a team, if we believe that those we work with are smart, good-hearted, hardworking, and an overall asset to the project, we will be much more likely to share information, provide them with what they need—including our time—and, in general, act in a magnanimous way toward them. But if, on the other hand, we perceive them as lazy, selfish, stupid, or a threat to our well-being, we will likely act in a guarded, suspicious, and otherwise adversarial manner. In either case, whether we choose to act magnanimously or as adversaries, we will feel justified in doing so. But the question then is: How reliable are the data that we are using to anchor these justifications? As Seneca said nearly two thousand years ago: "The wise do not put wrong construction on everything."

Yet, it is often the case that some very onerous and critical conclusions are arrived at based on rather limited samples of employees' behaviors, blended with extrapolations from past experience—that may or may not be relevant. How does this impact our teams? If we jump to negative conclusions about the motivations of others, we can create massive snags in workflow before a project has even gotten started. The reason this is so is because how we act toward others often dictates what we will get back in kind.

Our attitude reflects the regard we have for others. It is the truth that lies beneath the mask of all of those behaviors that we acquire at leadership seminars. The following paragraph from the Arbinger Institute's *Leadership and Self-Deception* says it best:

> The point is that we can sense how others are feeling toward us. Given a little time, we can always tell when we are being coped with, manipulated or outsmarted. We can always detect the hypocrisy. We can always feel the blame concealed beneath veneers of niceness. And we typically resent it. It won't matter if the other person tries managing by walking around, sitting on the edge of the chair to practice active listening, inquiring about family members in order to show interest, or using any other skill learned in order to be more effective. What we'll know and respond to is how that person is *regarding* us when doing those things. (2000, 27)

I've witnessed the veracity of this statement play out many times. Early on in my career, I was asked to assess two projects simultaneously and was told that both were suffering from the same malady. "Both of these project managers are living in the Stone Age," the operations manager said. "They think the best way to get things done is to yell and scream at people."

At first glance, the OM was correct; both PMs were relying heavily on their tempers to get things done. And the staff on both projects had plenty of tales to tell of being dressed down both publicly and privately—and neither liked it one bit. But after digging deeper into each situation, the data led me to very different conclusions and recommendations. For one, I suggested an intensive team-building session, with individual coaching for the PM. For the other, I recommended the PM be replaced, and perhaps dismissed from the company. Now you might well ask: Why would that be? After all, both project managers were displaying the same behaviors—yelling and screaming to get their way. So, why would I be soft on one and hard on the other? The answer rested on *how* each team described the behaviors in question—and the underlying regard that each of these descriptions revealed.

In the case of the first PM, people had this to say: "Yeah, he's a yeller and a screamer. But when he yells, he is usually right and he knows that you could have done better because he knows what you are capable of." Though he was extremely harsh when he perceived that shortcuts were being taken, he never beat people up when they had tried their best but simply made the wrong decision. His primary attribution error was assuming that people knew more than they actually did, mistaking legitimate confusion for lack of effort. But he more than made up for this shortcoming by being fully engaged with his team. It was clear that he had taken the time to get to know his people and had a strong grasp of each person's strengths and weaknesses. He also knew whether or not they were married, had kids, or if anyone on the team had health problems that might possibly put them off their game. And he wasn't at all shy about giving praise when they had turned their performance around or fighting for well-deserved raises for them—even if it meant raising the ire of his superiors.

Now, let's turn our attention to the second PM—or as his team nicknamed him, "Little Hitler." People on this project recounted instance after instance of his flying off the handle, accusing them of malfeasance whenever the least thing went wrong. On the rare occasions when the inaccuracies of his initial reaction were pointed out to him, rather than admitting to them, he often justified his behavior on the basis of some rather vague and spurious complaints that he ascribed to the owner. Rarely did he ever acknowledge positive performance, and when he did, he insinuated that their success was attributable to something he himself had done. He was so disengaged from his team that when he took his first job walk, nearly nine months into the project, the security guard attempted to escort him off the job site because he failed to recognize him. All Little Hitler cared about, according to his staff, was making sure that his own bonus was as large as possible.

Do you see what I'm driving at here? There was a reason why the staff continued to follow the first PM despite some of his inimical behaviors, while those under Little Hitler, as one person put it, "wouldn't have bothered urinating on him if we saw him on fire on the side of the road." The first team didn't like being yelled at, but they got that the PM cared—his high regard for them still showed through. Those under Little Hitler also got it. They saw that the only regard he had for them was for how they made him look. Things on his team had deteriorated to such an extent that people were actually celebrating failed deadlines in the hope of making him look bad. And there is no more deadly place for a leader to be than this—hence the difference in my recommendations.

Our regard for others can be revealed by something as simple as the questions we ask. For example, let's say that one of your young subordinates called for an inspection in an area that wasn't ready, and as a consequence, you received an earful from the inspector about how one of your staff just wasted his valuable time. When you inevitably call this young person into your office, you have a critical choice to make. You could ask: "Tell me what you were thinking when you called for that inspection?" or "What the hell were you thinking?"

First, let's be very clear about something. Having regard for others doesn't mean ignoring problems or not pointing out someone's shortcomings. In fact, failing to do so actually demonstrates a lack of regard by revealing that you believe the person who is underperforming has little capacity to improve. But *how* you draw attention to the issue is the difference between Lean thinking and cutting someone off at the knees. Toyota, the company at the forefront of Lean, believes that deep reflection, or *Hansei*, is the cornerstone of continuous improvement as well as creative problem solving. Or as George Yamashina, president of the Toyota Technical Center, explains,

> *Hansei* is a mindset, an attitude. At first, you must feel really, really sad. Then you must create a future plan to solve that problem and you must sincerely believe you will never make this type of mistake again. (*The Toyota Way*, 257)

But this mindset isn't achieved through yelling and screaming. Nor is it achieved by ignoring problems. It is attained by inviting people to engage in thoughtful inquiry. As Andy Lund, a program manager for the Toyota Siena, points out,

> People ... may not understand that it is not the objective to hurt the individual, but to help that individual improve—not to hurt the program but to show flaws to improve the next program. If you understand that deeply, you can get through that constructive criticism. (*The Toyota Way*, 259)

At Toyota, thoughtful reflection is achieved by focusing on the process, not results. They found that when managers focused purely on results and those were not attained, they tended to assume that a person or persons were to blame, and efforts toward continuous improvement stopped there. But when managers focused on the process and asked people to reflect on what caused the failure, they were much more likely to accept responsibility for the role they played in the failure and identify ways to prevent it from happening in the future.

So, going back to our example and viewing it with the concept of *Hansei* in mind, the first question is clearly an invitation to engage in reflection. By asking, "Tell me what you were thinking when you called for that inspection?" the person is given the opportunity to reflect on process issues, such as (1) timing (assumed that the area would be ready and attempted to save time by requesting the inspection too early), (2) execution (planned to cancel the inspection but were diverted by a competing issue), or (3) planning and judgment (due to inexperience they believed that the area in question *was* ready for inspection when it wasn't). The value of taking this approach is that it allows others to reflect on the root cause of failures more objectively, thus setting in motion a thought process to help prevent similar errors from occurring in the future. It also affords you the opportunity to provide strategic coaching and training with the same aim in mind.

Conversely, the only response that the question "What were you thinking?" evokes is defensiveness. Rather than engaging in productive reflection, the person in the glare of such an accusatory spotlight will either spin yarns or shut down entirely. The only lesson he or she will ultimately learn is how to accept a tongue lashing or how to avoid taking any initiative or additional responsibility in the future.

In terms of attitude, there is another critical element in play ragarding trust. By asking the employee, "Tell me what you were thinking …," you are actually giving the person who made the error the benefit of the doubt—that they may have had the best of intentions, but that their thought process was simply a little off—something that is both understandable and correctable. But by asking, "What the hell were you thinking?" the leader actually reveals an underlying contempt. Such a declaration signals an assumption that the subordinate's intentions are indeed suspect by implying that the failure was attributable to his or her being either stupid, lazy, incompetent, or intentionally careless—without actually saying so directly. But believe me, though not stated overtly, the subordinate will pick up on the underlying message—and he or she will resent it. But the real tragedy of such an exchange is this: instead of focusing on process issues, such as timing, execution, or planning and judgment errors—things that are correctable—the manager instead begins to focus on individual attributes—something subordinates can do little to fix. After all, we can't readily fix character or personality issues. But we can correct execution, timing, and skill related errors, which is precisely where the focus should remain. Assuming the worst about the intentions of others actually paints leaders into a corner: the only "fix" that can happen when we assume that people *are* the problem is to get rid of them!

This next point about attitude is more subtle still; trailers are very small places, and how we treat one person often conveys our regard for the entire team—whether we intend this or not.

After finishing an assessment of a team that was underperforming, I informed a soon-to-be crestfallen division manager that he was largely perceived as "intimidating and unapproachable" because of his occasional angry outbursts. He was incredulous. "How could that be?" he demanded. "My door is always open, and in all my years, I have never yelled at a single person in this office—ever!" He was being truthful: his door was usually open, and, to my knowledge, he never yelled at anyone on his team—at least not directly. But on the occasions when his door was closed, his staff overheard him scream at the top of his lungs at delinquent vendors and underperforming subconsultants. So even though his explosiveness was never directed toward them, these behind-closed-doors fits of rage conveyed to his team his *capacity* to devalue others. His team assumed that it was just a matter of time before he turned his rage toward them, so, preemptively, they did their best to avoid being the bearers of bad news and kept critical issues to themselves much longer than they should have.

The problem for most of us is that, in the moment, we are usually wholly unaware that we are demonstrating a lack of regard toward others. We think that we are just trying to get things done. We might have an inkling that we had done something to ruffle people's feathers, but we assume that if it were something truly off-putting, that someone on our team would let us know so we can make amends. Or if they remain silent, we assume that they have given us a pass because we achieved the results that were required. Let me disabuse you of this self-deception right now. Never count on getting such feedback directly or receiving any kind of pass. Just as no one will walk up to you, shake your hand, and say, "Another great job of leading me this week!" your underlings are equally unlikely to let you know when you have done something that has pierced them to the bone. The mere fact that you hold a title means that it is far more likely that your staff will tell you what they think you want to hear rather than risk telling, you something that might upset you. So, to increase your self-awareness, besides the obvious off-putting behaviors such as yelling and screaming, here is a list of the most common things that leaders do to inadvertently display low regard for others. They

- Don't return phone calls promptly (or at all)
- Show up late to staff meetings and expect everyone to stop and catch them up when they do arrive
- Chronically cancel staff or one-on-one meetings at the last minute
- Read or send text messages during meetings while others are speaking
- Avoid eye contact, fixating instead on their cell phones or computer screens
- Promise to provide a procedural tool or coaching and blow it off
- Neglect to follow through with addressing voiced concerns
- Respond to earnest questions sarcastically

I guarantee that if you do any of these things consistently, over time, you will successfully alienate your staff and foment the seeds of revolution. Worse, you will create an atmosphere diametrically opposed to Lean culture.

Some of you may be saying at this point, "Uh, I've actually done one or two of these things. Does this mean that I am a bad leader?" No! It means that you are a human being. In an environment as stressful as a construction site, it's difficult not to do any of the things from time to time. But if this is a consistent pattern of behavior for you, be aware that your chronic lack of regard will create perceptual voids that your team will fill in negatively. They will assume, often rightly, that there is always something of higher importance to you than respecting or responding effectively to their needs. You can protest to the contrary all you like, but your behavior will always speak louder than any of your words. What is under the mask will be revealed. Why is this important? Because any barrier that disrupts a smooth flow of communication, wastes people's time, or creates hesitation in others sets in motion the potential for lost productivity and diminished profit.

Let's go back to our kitchen example. If an inexperienced cook has a question about what the head chef meant when she wrote "fresh herbs" in the recipe, and the young cook holds on to the question because he perceives that, because of her abrupt attitude, the head chef is unapproachable, this increases the likelihood that a diner could get dill in her spaghetti sauce rather than basil—a rather bad outcome indeed. You could insist that the young chef in question "should have known better," but that is not the point. In any production process, whether it be an assembly line, a high-end kitchen, or a job site, errors such as these will continue to be repeated until either the customer complains or the person in charge happens to notice the mistake. This means that fully preventable mistakes will be sustained and repeated over a prolonged period of time until they are caught and stopped! With the right attitude exhibited by the leader—one where learning and continuous improvement are valued and honored—such errors are prevented before they have a chance to occur, because people simply aren't afraid to ask key questions.

Lean thinking is about taking every opportunity to address process issues early on and creating a culture that does not allow such problems to flourish. That's why I greatly admire managers like Terry Shugrue, a senior PM for Turner Construction in Eugene, Oregon. He refuses to allow such missteps to creep into the process—right down to how he schedules staff meetings. He carefully fixes a time for staff meetings each week, announces it to everyone verbally and via email, and publishes a corresponding agenda two days in advance. He then goes one step further; he copies top management and lets it be known that if they should schedule a corporate activity that conflicts with his meeting, that neither he nor any of his staff will be in attendance. He values the opportunity the team has each week to review the overall project plan and put their worries and concerns on the table, and he isn't about to pass it up for anything—even the opportunity to make himself look good by attending a corporate function.

When lack of regard is the root cause of problems, this next issue often takes center stage. It is a dynamic that plagues restaurants and job sites alike: when someone with unique technical skills is allowed to run roughshod over the rest of the team. Anthony Bourdain cites example after example of cooks he had hired who possessed extraordinary technical skills, but whose distinct lack of regard for their teammates (they showed up late, talked behind people's backs, didn't live up to team commitments) either killed overall productivity or drove others to quit. Every time he chose to look the other way, waste increased, the restaurant lost money, and Bourdain felt personally burned. He came to realize that by deciding to overvalue technical prowess, he was demonstrating low regard for the rest of the team by in effect saying: "Forget about what I said about the importance of us working together as a team, because what I really value is lone-star talent."

This is no different in construction. Yes, the owner may love a particular engineer for his or her ability to crank out PCOs or run an owner/architect/contractor meeting, but if, within the team, he or she withholds information, bad-mouths field counterparts, publicly belittles the administrative assistant, or in general, shows more interest in his or her own rising star than what is being produced as a team—and if you fail to address it—you are, in essence, telling people that the regard you have for teamwork is pure lip service. Trust me on this: I've seen many a team with a "star" technical performer go down in flames because their teammates viewed their behavior as so objectionable that they went out of their way to avoid or sabotage them. Instead of planning and coordinating, they erected fences, branded their cattle, and let the chips fall where they may—which often resulted in needless write-downs and angry owners.

Conversely, I've seen leaders with solid but not extraordinary technical skills do phenomenal things with inexperienced people—simply because of the high regard that they demonstrated toward them. In one particular job, a general contractor landed a project in eastern Washington constructing a new state prison facility. In retrospect, it was a job that the company probably should have passed up. It was a design/build job in a remote area, with unfamiliar subcontractors, a prohibitive contract, and an owner that, ten years prior, they had ended up in litigation against. To top things off, a substantial number of those on the job were fresh out of college. But as often happens, the market was dry, and in their desire to retain as many good people as possible, top management took a gamble.

When I went out to assess the team and heard what they were up against, uncharacteristically, I assumed the worst. I expected morale to be in the tank. The real question on my mind was how much money the company would have to write down—and how many good people they would lose in the process.

When the survey results came back I was stunned. They revealed surprisingly robust numbers in the areas of top-to-bottom and bottom-to-top communication, better than average role clarification, solid ratings in workflow and execution, and surprisingly high morale. How could this be? I knew that the project manager and the project superintendent were more than competent, but they weren't exactly human dynamos in terms of their technical prowess. And while each had ample construction experience, both were fairly new to their respective leadership roles. So how could I account for the fact that this team was more than holding its own—and against some pretty steep odds? Here is what stood out: the attitudes of the project executive (Jim Goldman), project manager (Eric Wildt), and project superintendent (Dwayne Goddard) were extraordinary. From the beginning, they made a commitment that, despite all the problems with the job, they would create an atmosphere where people could feel good about giving their very best each and every day.

The goals that they set for themselves was fairly straightforward: to be highly accessible, to encourage people's questions, to answer those questions (or pull in the appropriate resources to do so), and to pass on as much knowledge as they could. They also put in the time and effort to create an organizational structure whereby everyone on the team understood not only their own job responsibilities but those of their teammates as well. In their minds, the difficulties of the project were not excuses for people not to be prepared to take on the increased responsibilities that their next role assignment would demand. This is the very essence of responsible leadership.

Needless to say, this is pretty heady stuff. I've heard such lofty ideals expressed by management teams before, but I've rarely seen them implemented. But implement them they did—and it worked. "Do you know the greatest thing about this job?" many of the staff said. "If you fall down, there is always someone there to help you get back up again—not to do your job for you, but to help you pick yourself up. It's an attitude that everybody has around here. We do whatever we can to help each other—and it starts with those guys."

Think about how powerful this statement is from a Lean perspective. High standards were expected, but no one was expected to be on an island; the management team saw it as their duty to help people succeed. When mistakes were made or people had questions—and there were plenty of both—the managers put aside what they were doing, rolled up their sleeves, and simply asked, "What can I do to help?" As a result, instead of burying mistakes, people actually stepped up and took responsibility for them. The management team didn't have to turn over rocks to get to the truth because it was provided up front. More importantly, instead of becoming frustrated and impulsively taking things over when things went wrong, this management team was actually teaching people how to identify and fix their own mistakes so workflow would be minimally impacted in the future.

This job took a ton of work and required intensive support from top management to help right the budget, but against everyone's predictions (including my own), the project finished just slightly behind schedule and managed to eke out a small profit. But I am certain to this day that if the management team had been comprised of leaders who valued technical prowess over teamwork, the outcome for this project would not have been so rosey.

At the risk of being perceived as a complete Pollyanna, I want to share one more story about this project. As I mentioned before, this was a design/build job—but for much of the time, construction outpaced the often overmatched design team, and the job quickly became build/design—with all the inherent frustrations, reworks, and costs that this tongue-in-cheek moniker implies. But the day did arrive when design was substantially complete and the team needed to shift gears. Unfortunately, by then, the team had become so used to thinking "design first" that they repeatedly missed opportunities to execute the schedule more aggressively. The project manager (PM), project superintendent (PS), and project executive (PX) also recognized that the staff had grown complacent about underperformance by some of the subcontractors who also were suffering from this same design-first mindset. Therefore, the management team called a meeting to address the problem, but they did so in a decidedly *un*-heavy-handed way. PM Eric Wildt calmly came right to the point: "We have something very important to discuss. I know that we have to shift from being design driven to being schedule driven, but, to be honest, I'm not sure of the best way of getting there. What do you all think?"

Please take a moment to fully appreciate the brilliance behind this tactic. Surely, Eric and the rest of the management team knew full well how to go about making this shift and could easily have issued a series of edicts to simply make it happen. And if a management team is purely results driven, this is exactly what they will do: put out an edict, punish those that fail to comply, and keep pushing until they get the result they want—regardless of the mistakes and carnage that crop up along the way. But these managers realized that if they followed the usual path, they would be missing out on a golden opportunity to assess several important aspects of current team functionality that could help them improve the entire process for the long haul. For instance: Do these young folks know what being schedule driven actually means in the day to day? Do they understand the tools that are in place to help make this happen? And, if they know what it means and understand the tools, do they have the skills to make it happen?

By asking the question in the manner that he did, Eric also conveyed something that was a huge confidence boost for this young team:

regardless of their inexperience, the managers held them in high enough regard to ask them to generate a possible solution. This is the very essence of what Toyota refers to as *Kaizen* (meaning continuous improvement)—that the people who do the actual work should be involved in all continuous improvement efforts—regardless of their experience level.

And through his willingness to express vulnerability and admit that he wasn't exactly sure of the best way for this particular team to make the shift, the PM gave the rest of the team permission to admit the things that they didn't know, which turned out to be quite a lot. For instance, while many people knew that a schedule existed, very few of them actually knew how to read it. Even those who did know how to read it weren't clear about how to use it as a guide for planned actions. And this was just the tip of the iceberg. After further discussion the team generated the following plan of action:

> **Live the schedule: Clear the path and take away subcontractors' excuses for not performing.**
> **Goal: To become proactive rather than reactive.**

We, as a team, will not allow this goal to fail. Therefore, we commit to the following:

- We will all attend a workshop hosted by the PS to learn how to read the schedule.
- We will read, reread, and ask questions about the schedule until it makes sense and becomes second nature.
- We will shift our staff meetings' focuses from design issues to schedule-driven issues.
- We will audit our submittals.
- We will walk the field as engineer/superintendent partners, hand carrying copies of the schedule.
- We will generate engineering "hot lists" and conduct daily, ten-minute "What's hot/what am I worried about?" meetings, either at the beginning or end of the day.
- If our best efforts to improve performance fail, we will call on our teammates for their assistance.
- If we ourselves are in error, we will own it and rectify the situation.
- And most importantly, *we will change our attitude* toward nonperformance. We will do our due diligence, but we will stop doing other people's work for them. We will insist on others doing the jobs that they are paid to do. We will act like responsible adults and will expect others to do the same!

Not bad for a bunch of rookies! It did take two hours of discussion to generate this plan, but also by the end of the meeting, everyone not only understood it, but bought into it as well. They did so for one simple reason: because they helped create it. Weigh that against the ostensible expediency of issuing edicts that are only fully understood by 20% of the staff, and you can easily calculate the cost of taking such a shortsighted approach.

When you are attempting to create a Lean culture, it's about viewing the job in a broader context and setting the team up to achieve sustainable results over the long haul, rather than implementing quick fixes that could, in the long run, be fraught with problems and misfires. Despite all the engineering techniques that permeate this burgeoning modality, Lean construction is, above all else, a people-driven concept. This simply means that we must to pay attention to the how's of what we do. To do this, we need to take the time to treat people like human beings who have a brain between their ears, as opposed to shortcutting the process and leaving them feeling like mere extensions of their laptops or tool belts. The more we treat people like human beings, the greater the likelihood that they will stay engaged and invested in keeping the assembly line moving in the most efficient ways possible.

By extension, this same principle applies when we need to do the hard things, such as confronting people about their poor performance. If our aim is to sincerely help someone to improve—rather than merely inflicting a wound because we feel irritated and therefore justified in doing so—we should be able to deliver a message that leaves employees feeling like they can make the needed improvements versus feeling that they have just been ground into the dirt for personal reasons.

I hope that you are beginning to see that creating a Lean culture doesn't require you to learn a plethora of leadership techniques or to undergo a total personality overhaul. It simply obliges you to gain an increased awareness about the impacts that your attitude—both positive and negative—can have on others. This also includes how you approach the construction documents that you have inherited from your preconstruction team. No matter how terrible you think the contracts or drawings are, own them! Great management teams don't waste time crying, "Woe is me." Complaining about your documents does nothing but show your low regard for the top management team that put them together—and encourages underlings to do the same. Worse, it gives your team a built-in excuse for failure. The message you need to give your team is clear and simple: these are our documents and it is our job to execute them to the best of our ability every single day. When you think about it, what more can you ask of your team—or of yourself?

This begs another important question: How do you regard yourself? How badly do you beat yourself up when you make a mistake? Being responsible for your actions is great, but brutalizing yourself over errors isn't—and will probably lead you to commit a number of Lean-killing behaviors, such as (1) being even more impatient or intolerant of the failings of others (and feeling justified in doing so), (2) taking on even more responsibility as a form of twisted, self-punishment. Most managers that I know beat themselves up ten times worse than they do the people around them. Lighten up for gosh sakes! All you can do is your best—so have some regard for yourself as a human being! If you are able to view your own mistakes more objectively and less harshly, the rest of the team will benefit from this healthier means of self-reflection as well! As Seneca said,

> I'll tell you what took my fancy in the writings of Hecato today. "What progress have I made? I am beginning to be my own friend." That is progress indeed. Such a person will never be alone, and you may be sure he is a friend of all. (*The Consolations of Philosophy,* 103)

One last story, and then we'll wrap up this section. This is important because some of you may be called upon to replace a manager who has botched things pretty badly, and your attitude will be the key in turning such a difficult situation around.

Scott Miller (PM) and Dennis Newman (PS) were brought in to clean up a job in San Francisco that was, in a word, a mess. The start of the job couldn't have gone more poorly. Soon after breaking ground, the team hit human remains, Native American artifacts, and an unknown underground stream. And under the terms of the contract, the GC "owned" all of these site conditions—and the number they had in their estimate for these unknown conditions didn't cover it. To make matters worse, the original PM and PS were reactive versus Lean thinkers. Planning was a complete afterthought as they ran from one fire to the next. They doled out tasks as they became critical, rather than in accordance with an overarching plan. While everyone on the team was working extremely hard, you'd never have known it by their work product. Their efforts resembled that of a crew on a sinking ship. While each of them frantically bailed as hard as they could, because there was no fully integrated plan, all they managed to do was fruitlessly take their buckets of water and pour them directly into the compartments of those standing along side of them. Morale, predictably, was in the Port-o-Potty, and the owner was absolutely livid with the GC's performance.

This is the point where Scott and Dennis stepped in. It certainly didn't hurt that they had fifty years of construction experience between them, but keep in mind, so did the people that they replaced. Now, they could have come on the scene and said, "Well, we heard that you people really screwed this job up, and we're here to fix it," but they didn't. Instead, they introduced themselves, had everyone else do the same, and then announced, "We know that you folks have been up against it for quite a while, and we're here to help." Over the next few days, Scott and Dennis did the following:

- Via individual interviews, they found out what each person had been doing, what their experience levels were, and, from their perspective, what had and hadn't gone well.
- As managers, they made a point of staying out of their offices. They met around a formerly unused table that was originally meant for staff meetings, poured over drawings, and pulled others into their discussions to ask their opinions.
- Whenever someone on the team looked down or fearful, they reassured them that though the project was tough, they'd been through tougher—and they believed that this team was capable of pulling this job off.
- They produced a new organizational chart that created assignments commensurate with each person's abilities, reviewed it with everyone, and collectively went over everyone's roles and responsibilities.
- They invited people's questions and input, made themselves available to answer questions, and provided guidance and training whenever needed.

When I came back three months later, I was amazed to see the turnaround in the staff's confidence. Equally impressive was the quality of their work product; a job that had been utterly stagnant was now emerging rapidly out of the ground.

Unfortunately, I can't relay a fairytale ending for this one—the project did lose $3 million. But defying all expectations, it finished on time and received favorable recommendations not only from the architect and engineer (A&E), but from the owner as well—saving untold millions in potential future earnings. And there is no telling how much more money would have been lost if Scott and Dennis hadn't come on board.

Here is the model I penned to capture what Scott and Dennis did to such great effect; it puts a different spin on the notion of whipping people into shape:

WHIP-C Formula

W—Welcome. They viewed all as valued teammates rather than seeking out scapegoats or whipping boys.

H—Help. They shared knowledge, answered questions, and engaged in problem solving.

I—Invite. They invited participation, involvement, and input.

P—Participate. They modeled positive interactions and provided a sense of hope.

C—Clarify. They clarified roles, expectations, assignments, and the organizational structure.

5

Lean Ethics

In a word, each man is questioned by life; and he can only answer to life by answering for his own life; to life he can only respond be being responsible

Viktor E. Frankl

If our attitude is to remain consistent, it must be grounded in something more enduring than the ever-present impulse to get the job done. If getting the job done is our only desire, then we could be tempted to take questionable shortcuts in our leadership practice that are decidedly *not* Lean. Therefore, we need to look beyond emotion toward a broader base of sustainable wisdom—our ethics.

Why are our ethics germane to our discussion of attitude and developing a Lean culture? Because, when distilled to its essence, leadership it is both a virtue and a sacred trust. Whether you have been selected to be a general manager, operations manager, regional or division manager, project executive, project manager, project superintendent, department head, lead engineer, or general foreman, you need to recognize the honor that has been bestowed upon you. These titles are not given to just anyone who happens to come along. It means that people within your organization with a proven track record of experience, know-how, and success think highly enough of your ability and character to entrust critical objectives to your judgment. Even if, at this point in your career, you believe that you lack some of the necessary skills or essential qualities to be a skillful leader, it is important for you to trust in their belief in you.

So, does this mean that the people who work under your direction will suddenly do everything that you ask simply because some fancy letters now appear after your name? No, of course not! That's why it is imperative to adopt a way of thinking that taps into a source of wisdom that is much more substantial than our own ego-driven desires.

ETHICS

In modern times, we tend to limit our examination of ethics to a litmus test as measured against a legal standard. For instance, we may ask, is it ethical to accept gifts from subcontractors with whom we may be doing business with in the future, and could this be construed as biasing our decisions in their favor should their selection be challenged in court? Or, is it ethical for a general contractor to make a large donation to a political candidate when they may be bidding on projects within this politician's sphere of influence? Such examinations are specifically aimed at keeping companies out of legal hot water. But in reality, such questions are rather limited in scope.

The ancient philosophers, typified by the likes of Socrates and Seneca, didn't view ethics as a test to be narrowly applied to specific cases in order to prevent potential legal consequences. Their scope of ethical consideration was much broader. They viewed ethics as a guide, a means toward living a good and purposeful life—something to be reflected upon and practiced each day. If they were able to travel forward in time, they would be puzzled by our response to their query as to whether or not we thought about ethics. We'd likely say, "Yes, I believe we have an ethics department at the local university that specializes in that sort of thing." "Really?" they'd likely respond. "Do your universities have 'love departments' and 'friendship departments' as well? Is the university the only places where such important matters are considered?"

So central were ethics to "living the good life" that Socrates went so far as to posit that anyone who focused on anything other than what was right and good (albeit money, ambition, power, or prestige) was clearly insane—because the pursuit of all of these things for their own sake ultimately led to unhappiness. For Socrates, living a good and ethical life was the only clear path to true happiness.

In this wider view, ethics are composed of three elements:

1. **The good**—the thing desired, the ideal. It is at the heart of what we mean when we wistfully declare that someone is a good leader. In the construction world this comprises all the technical and artful things that a leader does in an ideal way (i.e., thoroughly reviewing and vetting scopes of work and contract documents, understanding their intent, and conveying this information to the staff). Your company core values should be also reflected in your pursuit of the good.

2. **The right**—the opposite of wrong as defined by law. Embezzlement, lying, taking money from petty cash for your personal use—these are always wrong. Complying with sanctioned codes, honoring bids according to the rules of requests for proposals, not swapping personnel who were originally proposed and promised, and maintaining safety practices are also part of the right.

3. **The ought**—personal obligation, duty, responsibility. Providing training and coaching, giving accurate and timely feedback about performance, sharing information and knowledge, and otherwise making investments in your teammates' success—even if, personally, we don't like some of them—are all implied leadership oughts.

C. S. Lewis, a brilliant philosopher who helped maintain his countrymen's morale as England was getting its collective buttocks handed to it by Nazi Germany at the start of WWII, often used the following analogy to describe the importance of ethics in day-to-day leadership. He instructed leaders to imagine those working under their direction as individual ships docked in a harbor. As they bob up and down, each of these expectant ships is waiting for something critical from their admiral—their sailing orders. For our purposes, whether they be estimating, purchasing, engineering, or field ships, they are all waiting for the leader to give the order to commence. These orders, according to Lewis, are comprised of three essential elements:

1. *How to cooperate and coordinate with one another so the ships don't collide.* Otherwise known as *social ethics*, they comprise the understandings that each person has of one another's roles and responsibilities, as well as the responsibilities that each "ship" carries in terms of communicating their position (i.e., where they are, what they will need, and when they will need it by), in order to maintain their position in support of the overall fleet. They also establish the order by which the ships will leave the harbor. (The engineering ship is always ahead of the field ship).

2. *How to keep each ship afloat and in good condition.* These are *individual* ethics or *virtue* ethics. They constitute a thorough understanding and complete commitment to keeping one's ship in good working order and executing a particular role within the fleet. In the construction world, each person on the team needs to be provided a clear set of expectations so they can accept full responsibility for

executing said role to the best of their abilities. In Lean terms, this means that each person is tasked with developing and maintaining a work plan that is congruent to the overall project plan.

3. *What is the ship's mission?* In other words, where are we going as a team? This is the most important order of all because it gives everyone on the team a target in terms of their ultimate destination in accordance with the overarching plan. As the Mad Hatter once told Alice, "If we don't know or care where we are going, it doesn't make much difference how we get there, does it?" For our purposes, we need to ask ourselves, as leaders: Does everyone on the team understand where the final destination is (end date)? Have we laid out the logic of the job for getting there? Have we been able to transmit what we know about the scopes, plans and specs, and expectations in a manner that everyone can understand? Will my teammates be able to comprehend the important mile markers that indicate that we are on or off course and be able to make the necessary course corrections if needed?

I hope that, as you read the above description, it became apparent that the only person that your people can reasonably depend on for getting these important sailing orders is you—the leader! After all, you are the one who interprets the contract, sets the course, and creates the vision so that the team can reach its final destination. But, having said this, let's shift gears for a moment.

So far in this book, we've taken a philosophical position that views mankind fairly optimistically, that is, that people want to do good and, with a little help, direction, guidance, and clarity from their leader, they will. But if you've been around the block a few times, you know darn well that mankind is *not* comprised entirely of those whose motives are pure, and a number of philosophers throughout history have put forth their own brand of recommended leadership practices to address this fact. Here is a sample:

- Men are so false, so insidious, so deceitful and cunning in their wiles, so avid in their own interest, and so oblivious to others' interests, that you cannot go wrong if you believe little and trust less. (Guicciardini, 1483–1540)
- If you are involved in important affairs, you must always hide failures and exaggerate successes. It is swindling but since your fate more often depends upon the opinion of others rather than on facts, it is a good idea to create the impression that things are going well. (Guicciardini, 1483–1540)

- It is much safer to be feared than loved. Love is sustained by a bond of gratitude which, because men are excessively self-interested, is broken whenever they see the chance to benefit themselves. But fear is sustained by a dread of punishment that is always effective. (Machiavelli, 1469–1527)

These quotes were compiled by a man named La Rochefoucauld in the seventeenth century as a guide—an employee handbook if you will—for those who were entering the French aristocratic court for the first time. They reflect a rather dismal view of the human condition. The assumption made is that people are so self-interested at their core that it is a waste of time to invite them to make sacrifices and in order to achieve some greater good, because at the first opportunity, they will turn things to their own advantage, often at our own expense. Therefore, they recommend that the best approach is not to engage others as willing teammates at all, but to preemptively manipulate them—or threaten them with punishment—in order to bend them to our will.

So, who is right? Should we model our leadership ethics after the likes of Socrates, C. S. Lewis, and other noble leaders and become the champion of the good and right—and invite others to join us in our quest? Or should we embrace the ugly truth espoused by Machiavelli, assume the worst, and prepare ourselves for the inevitable battle against self-interest that is sure to follow?

Surely, there is ample evidence throughout history for both viewpoints to hold water. As inspired as we are by the Mother Teresas of the world, we often feel downright foolish when attempting to follow their example, particularly in the workplace. After all, if mankind lacked a sinister side, we wouldn't need locks on our doors, antivirus software on our computers, or construction laws governing our contracts.

So where does this leave you as a construction leader? It would be preposterous for me to tell you that there aren't people out there who wouldn't gladly lie, cheat, and steal at the first opportunity. Indeed, there will be times when you will be called upon to confront those who are dishonest, deceitful, and greedy, or to defend your team from unwarranted attacks by the owner, or to address internal cancers who are threatening to tear your team apart. In fact, to *not* do so would be unethical. But here is the critical question: Is it truly wise and good to treat everyone, from the start, as if they were potential enemies of the state—particularly internally? Is it truly prudent to convey to the individuals on our team that they are not to be trusted until proven otherwise, or that their motivations are suspect every time they do something incorrectly?

Unfortunately, we are already more than primed physiologically to go in this direction from the outset. We are hardwired to detect threats everywhere—even for such seemingly innocuous things as informational voids—and to respond to them in kind. If you doubt the veracity of this statement, examine what you told yourself the last time you left an important message for someone higher up, and they didn't return your call. Did you assume that they were busy, had different priorities, or simply had something come up in their personal lives as the reason they didn't get back to you? Or did you become angry and convince yourself that their lack of response was intentional and that they were going out of their way to make your life miserable? If you are like the vast majority of us, you probably assumed the latter.

This means that as leaders, we have to be aware that we all are physiologically biased toward negative assumptions about the people we work with. This means, if you are to be effective, you have to be diligent about keeping these unhelpful biases at bay. The difficulty we face is that we tend to scan for confirmatory evidence (in the form of negative behavior) to substantiate our initial beliefs as a means of keeping ourselves protected from potential future harm. Unconsciously, we look for data to confirm our beliefs, and don't pay attention to that which contradicts it. In fact, when examining leadership failures across disciplines, be they in the realm of politics, military combat, or job sites, we find that they tend to center on such fundamental errors in critical thinking. Leaders unwittingly select their audience, listening to those who already agree with their biased positions, and dismissing those who do not. (JFK was so convinced of the rightness of his plan to invade Cuba that he selectively gave more weight to the advisors who backed his plan versus those who gave more than ample evidence to the contrary.)

As leaders, we also must recognize that we are not alone in this distorted way of thinking. The people we work with, internally and externally, struggle with these very same cognitive biases. Perhaps this helps explain why, when you slid the first change order onto the owner's desk, he or she angrily exclaimed, "I knew you were going to change order me to death!"—without even so much as examining the validity of the claim.

If they are so detrimental to human interactions on all levels, you may wonder why it is that we so readily adopt biases in the first place. The reason for this is largely biological. Our brains are literally organized for organization. We like order. We are wired to look for patterns and align them into meaningful patterns. In fact, your very success as a construction professional is dependent upon this hardwired tendency. It is the reason why when you look at a set of blueprints, you are able to discern meaning out of all those squiggly lines and do something useful with the information. And we also do the same thing when we think we detect behavioral patterns.

We adopt biases because they are an easy shorthand for sorting out the complex world around us. If we didn't, we simply couldn't survive the onslaught of data that flood into our trailers. On any given day, the average project manager interacts with scores of people—owners, architects, consultants, city inspectors, subcontractors, and lawyers, to name just a few. How do we sort through the massive information overload that bombards us each day? Unconsciously, we look for behavioral patterns that we believe "sum up" the intentions that each person is presenting to us. The whole notion of intuition or gut feeling is nothing more and nothing less than this. We simultaneously learn and feel our way through situations—and remember what we have learned as intuition. This is actually what we mean when we say that someone is experienced, that they have acquired a knowledge base by having dealt with similar situations to those that will be presented with in the future. Unfortunately, along with this knowledge, they will also have acquired inaccurate biases along the way as well. Believe it or not, intuition is a skill like any other. And like such things as driving a car or creating a schedule, some of us are better at it than others. But unfortunately, this doesn't stop us from adhering to our biases once they are formed. We actively maintain our biases because we *believe* that they will somehow protect us from harm—and because sometimes they actually do. We can all recollect times when we've been able to head off a claim or detect an inflated estimate because we were able to feel it in our gut that something was off and act on our skepticism. So why then shouldn't we, like Machiavelli, assume the worst about others as our default position and constantly scan for nefarious intentions? How could we possibly go wrong? By assuming the worst we can never be taken advantage of, right? At least that's the belief anyway. But is it really true?

It's been my experience that assuming the worst about others creates more problems than it solves. In order to fully assume the worst about someone else, we need to inflate all of their faults and, in doing so, must necessarily inflate our own virtues—which necessarily distorts the objective reality of the situation. If you doubt this corollary, examine your own thought process the last time you felt that a boss came come down on you unfairly. It probably sounded something like this: "He's always such an unreasonable jerk! How can he be that way toward me? Can't he see how hardworking I am and how I always give my best?" This is what we call in the psychology business a cognitive distortion. It is the process by which we systematically distort our view of others and ourselves. And once we go down this road, objectivity flies right out the window, and so does our ability to reach mutually acceptable agreements. The fact is, no one is always unreasonable, and nor do we always give our best. But that is not how we see things when emotion takes over. Instead, we lock into our positions, act badly toward others, and feel completely justified in doing so. And all this does is invite others to do exactly the same. At this point, we stop looking for ways to help things go right, and instead seek out justifications for continuing to act in decidedly unhelpful, unyielding, and ultimately, in ethical ways.

So, how can we avoid this trap? One way is to recognize that it is our ethical responsibility as leaders to address our cognitive distortions for one explicit reason—so we can keep the job moving effectively. Here is an example: Let's say that you are a project manager and before heading to your office you decided to walk the site and check out an area where a critical concrete pour should be occurring. But when you reach the appointed spot nothing is happening. In fact, there isn't a single cement mixer to be seen anywhere. Even worse, your superintendent, the one who you discussed this issue with just the day before, is also nowhere in sight! To be honest, you've been frustrated with the superintendent's closed-mouthed approach to his work for quite a while, but this goes beyond the beyond! "Who does this guy think he is?" you tell yourself. "Am I going to have to do his job as well as my own?" If uninterrupted, this course of thinking will lead you down a path that will disrupt workflow even further. For instance, since you believe the issue was already thoroughly discussed the day before, it is an easy leap to assume that it is up to the superintendent to approach you about what went wrong—not the other way around. So you sit and wait, and as you do so, you stew in your own juices. By the time the superintendent enters the trailer you are truly steamed. When he looks over to your office you don't say a thing—you just sit and glare. And when he does try to speak, you make a sarcastic remark about how nice it is that he has finally shown up. As a result, the superintendent, in turn, doesn't say a word. He just angrily turns and walks out of the trailer. Let's examine all this in tabular form.

COGNITIVE DISTORTIONS

Cognitive distortions are things that we tell ourselves that push us away from doing what we need to do, or to justify why we won't or can't do what we should do to help others (Figure 5.1).

It is important to recognize that the way we think about things—what we tell ourselves in the heat of the moment—can have a profound impact on our attitude, our emotions, and our subsequent actions. By convincing ourselves that we have been wronged, we not only commit to an adversarial course of actions, but also we become further convinced of our rightness and the other person's wrongness.

To head this off, we need to insert a process that interrupts this cycle and allows us to remain on an ethical path. We need to invite ourselves to engage in rational debate that can alter a potentially negative outcome (Figure 5.2).

When we are able to talk ourselves down from an emotional ledge by injecting a healthy dose of reason, we give ourselves the opportunity to minimize deeper disruptions to workflow. In this case, by remaining calm, the PM would have discovered that the superintendent had actually made

Incident	What I Told Myself	Result of Distortion	Debate
Walk the site: Notice that work isn't occurring in areas where you believe it should be (or as was discussed with the superintendent just the day before).	"That's not my job! That's the superintendent's job. I've got enough on my plate—I shouldn't have to do his or her work too! And we just talked about this!"	Don't talk directly to the superintendent; instead, give him the cold shoulder and make a sarcastic comment when he does speak. The superintendent doesn't get the message, stops talking altogether, and storms out of the trailer. Nothing, as far as you can tell, changes in the field.	

FIGURE 5.1
Tracking cognitive distortions.

Incident That Upset You	What I Told Myself	Result of Distortion	Debate
Walk the site: Notice that work isn't occurring in areas where you believe it should be (or as was discussed with the superintendent just the day before).	"That's not my job! That's the superintendent's job. I've got enough on my plate—I shouldn't have to do his or her work too! And we just talked about this!"	Don't talk directly to the superintendent; instead, give him the cold shoulder and make a sarcastic comment when he does speak. The superintendent doesn't get the message, stops talking altogether, and storms out of the trailer. Nothing, as far as you can tell, changes in the field.	"Calm down. We have different jobs, but this is 'our' project. If anything fails, we all fail. Maybe something came up that I didn't consider. I need to find a quiet place where we can talk privately and calmly ask him about what hapened."

FIGURE 5.2
Healthy debate.

a heady decision to cancel the pour after detecting that incorrect conduit had been put in place by the electrical contractor the night before. And at the time when he should have been overseeing the pour, the PS was actually in the electrical contractor's trailer, not only pointing out the problem, but also preemptively hammering out a recovery plan.

You could quibble that the superintendent should have immediately informed the PM of this fact, and that a simple phone call could have headed off a conflict between he and the PM—and you would be correct. But this dilutes the point. As a leader, it is important to remain on steady ethical ground. If we jump to conclusions and assume the worst about those we work with, we ourselves throw a monkey wrench into the machinery of the team process. In this example, the superintendent became defensive and self-protective in the face of the PM's seemingly unjustifiable "threatening" behavior. In essence, he felt punished for doing what he believed to be good and right by heading off a problem that would have negatively derailed the project to a much greater extent than a missed

Incident That Upset You	What I Told Myself	Result of Distortion	Debate

FIGURE 5.3
Blank tracking sheet.

pour date—thus violating his own ethics. In short, it is easy to see how cognitive distortions can quickly become Lean killers!

Figure 5.3 shows a blank form. The next time you are upset, examine whether or not you have fallen victim to your own distorted thinking, and see if you can find an alternative that can help you to maintain objectivity and keep the project on track.

Through the course of this book, extrapolations will be made, based on empirical studies, indicating how we, as humans, typically respond to positive or aversive stimuli. But does this mean that, in terms of our attitude and behavior, we are all slaves to our environment? Does it mean that what we feel and, consequently, how we act are purely determined by the situations that we find ourselves in—as if we have no will of our own? No! Regardless of the circumstances we find ourselves in, we can and should retain our ethics. In fact, this is the very essence of what it means to be ethical! Machiavelli once said that we only have to be honest or ethical if we perceive that others are acting in this same manner toward us. But think about what this actually means. The reality is that Machiavelli had no ethical code; he just did, in kind, what he perceived others to be doing. That is *not* ethical behavior!

As leaders, it is our ethical responsibility to override our autonomic fight-or-flight responses to events that upset us. Viktor Frankl, a psychiatrist and concentration camp survivor, spoke to this key point in his seminal work, *Man's Search for Meaning*:

> Man can preserve a vestige of spiritual freedom, of independence of mind, even in such terrible conditions of psychic and physical stress.
>
> We who lived in concentration camps can remember the men who walked through huts comforting others, giving away their last piece of bread. They may have been few in number, but they offered sufficient proof that everything can be taken from a man but one thing; the last of the human freedoms—to choose one's attitude in any given set of circumstances, to choose one's own way. (1959, 75)

So, when the owner pressures you to shortcut your safety program in order to speed up the job, or when the owner's rep pushes you and your team well beyond their scope, or when your own stress response compels you to lash out at teammates, you do have a choice—and a responsibility—to rise above your biology and instead do what is right and good!

One final point, and I only add this because I see this type of situation with a fair amount of frequency. If you are the kind of leader who rewards, trains, and shares information based on who you think has your back, and feels free to withhold rewards, training, and information from those who don't, you are not an ethical leader. Though you may feel justified in conducting yourself in this manner, these are, in actuality, the actions of a Third World despot (i.e., if you take care of me, I'll take care of you; if you don't take care of me, watch out!). Though, sadly, much of the world operates in this manner, this is precisely the kind of thinking that contributes to so many of its ills. Over time, this thinking rapidly deteriorates to simplistic, black-and-white thinking—you are either for me or against me—and leads to work issues becoming overly personal. This way of thinking leads leaders to conclude that poor performance isn't just a work issue, but something personal, that the lack of performance was done to directly harm them. Worse, it leads leaders to believe that it is completely justifiable to treat others in harsh and inappropriate ways as some sort of wrong-headed retribution.

Ethical leadership is about adhering to a code that you apply equally to everyone, not just those who scratch your back. Truly ethical leaders aren't selective about who they choose to help, share information with, or support. They provide these things for everyone on the team—all for the sake of the greater good!

6

Construction 101: The Basics from a Lean Perspective

A lot of [Bigfoot's] time was spent figuring out ways to make the restaurant run more efficiently, more smoothly, faster and cheaper. And one of the earmarks of a Bigfoot operation is the tiny design features: the conveniently located hot-water hose for bartenders to melt down ice easily at the end of the night (into convenient drains, of course), the cute little plastic handles on any electrical plug near any station where a worker's hands might be wet. And everything is always easy to clean and easy to store. Pots hang from overhead racks, always in the same place. Bottles at the bar are arranged in mirror image, radiating out from a central cash register. Careful consideration is taken with every detail, from where employees store their shoes to custom-cut inserts for the steam table. (2000, 99)

—**Anthony Bourdain**, *Kitchen Confidential*

Let's shift our focus to the nuts and bolts aspects of leadership that can greatly impact workflow. As alluded to earlier, leaders can become so preoccupied with reacting to owner demands and pushing out deliverables that they can unwittingly worsen project flow by not attending to what I call the basics. By not taking the time to set up the job properly, they inadvertently prime it to be reactive. As people flail about in organizational uncertainty, they react to issues that come up as if they were wild fires, rather than taking the time to anticipate and plan for them. In the process, leaders find themselves pushing inadequately experienced or trained people to produce a product that is well beyond their capabilities. The result is an end product that is woefully lacking in quality, and an owner who is less than thrilled with the result.

So what are the basics, and why are they so important? In a nutshell, they are *organizational structure*, *flow*, and *feedback and positioning*. They are the critical means by which leaders establish the overall organizational design, set the ground rules for execution, and determine the sequential logic that everyone on the team can follow and rely upon. But before we discuss these key elements in depth, I want to put them in their proper context and, at the same time, challenge traditional thinking.

Often, construction managers look upon pricing, scheduling, PCO submittals, RFIs, budget reports, etc., as disparate activities—tasks that are necessary to complete, but are somewhat loosely connected to the overall project. To some degree this thinking is understandable. Compartmentalization is how we gain some semblance of control over the onslaught of tasks that must be dealt with on a day-to-day basis. It's how we manage to eat the entire elephant—one discrete bit at a time.

So here is the challenge: think of yourself as an overseer, not of discrete deliverables, but of the entire process—as if your project was one large assembly line in a humungous manufacturing plant. Your end date would represent the time when the fully assembled product would need to come off the line. Each milestone that precedes the end date would represent the critical subassemblies that go into the making of the final product. That means that each action that your team takes part in, be it the processing RFIs, PCOs, submittals, material handling, etc., should represent a significant contribution to these subassemblies. When done correctly, and in a coordinated manner, these actions not only keep the line moving, but contribute to the overall quality of the end product. On the flip side, if these tasks are completed incorrectly, or the timing is off, the final product will suffer. This means that any weaknesses in the design of the overall organizational structure that leads to confusion, poor coordination, or botched execution will result in half-backed subassemblies, which will, in turn, have a deleterious effect on the final product. The goal of every leader should be to anticipate such weaknesses in the structure, and create a design that promotes a seamless flow of communication up and down the line by setting the ground rules in terms of who does what, establishing the sequential flow of tasks and activities that need to be performed, and positioning themselves to ensure that there are no failure points anywhere along the way. If these key elements are in place at the beginning of the job, you'll have the best chance of delivering the finished product on time, on budget, and to design. If not, your project will struggle from preconstruction to its merciful end.

On its face, this probably seems laughably obvious. Yet, time and again, I've seen projects organized in such a manner that, if you didn't know better, you'd swear they were intentionally designed to incorporate these obvious flaws at their inception.

STRUCTURE

Eighty percent of the problem jobs I encounter are lacking in some or all of the following key structural elements. The bigger the job, the more vital these elements become and the more deleterious their effect if absent.

Organizational Structure

When managers are asked to create a project organizational chart, too often they look upon the assignment as one more unnecessary chore handed down by upper management. So they simply look at staff projections, count out how many engineers and field people they will be allocated, and create a chart based on of experience levels—filling in the less vital roles with whoever they have leftover. As new people come on board, they are slotted into conveniently created to be announced (TBA) to be announced vacancies. Managers who are a bit savvier obtain a rudimentary assessment of their underlings' capabilities from their operations manager or project executive and position people in accordance to their technical proficiencies. This is better, but it won't ensure that the structure will work. Again, you need to assess the structure as if looking at an assembly line. Here is the key question to ask: Have I created an organizational structure/chart that increases the likelihood of a smooth and steady workflow and minimizes the probability of inconvenient, structurally induced stoppages and waste?

Here's why this question is so important. Let's say you have a technical wonk on your team, a person so technically brilliant that he could recite the entire history of exterior enclosure systems from ancient times to the present. Let's also say that this person has the ability to generate estimates and buy out a job like nobody's business, and that he also knows your company's policies and procedures so well that he can easily write procedures manuals for most projects off the top of his head. Based on these skill sets alone, you'd probably be inclined to make this person a project manager or, at the very least, a senior project engineer, right? But what if I also told you that this person has a track record of being a very poor multitasker, that he tended to hoard information in order to make himself look good, and that on his last job, whenever there was a hiccup, he wrote scathing emails directed at the superintendent, which he habitually copied to everyone else on the team. Still tempted?

I can't underscore this point enough: Technical skills alone *do not* determine someone's fit within an organizational structure. Right fit is determined by both technical competence *and* a person's ability to keep the line moving via his or her attitude and interpersonal competence. To this end, here are some additional questions to ask:

- Do I have right people in the right positions based on both technical and interpersonal skills? (Having the wrong person in the right place or the right person in the wrong place will impact flow.)
- Do I have people in place who will require a significant amount of training? (If so, what is my plan to get them up to speed as soon as possible?)
- Are there any holes in the chart, that is, are there any missing players? (Will these holes cause people to drop down or across the organizational structure to fill voids?)
- Are there any inherent conflicts or bottlenecks that would disrupt flow? (That is, is there anyone who is performing multiple functions or who serves more than one master?)

Once you've sufficiently answered these questions, mentally test your chart. Do run-throughs for various processes and procedures and conduct hypothetical *what if* scenarios to make sure that there aren't any unforeseen bottlenecks.

One variation on this theme is an exercise that I call *org-chart-o-rama*. Simply clear out a conference room and ask people to assemble themselves as a living organizational chart. Off the bat, you'll be able to see if there are any potential misfires should people place themselves under the wrong person. Next, select a procedure and test it. For example: "In my hand is an owner-generated change order. Who gets it first, what do you do with it, and where does it go from there?" Go through this for as many steps as necessary and ask people to explain their rationale. This will help you to identify any misfires or confusion and give you the opportunity to provide your team with a tremendous amount of procedural clarity. It will also allow you to emphasize specific expectations for performance. For example,

> Okay folks. It's not sufficient for the engineers, particularly the lead engineer, to simply receive the change order from our administrative assistant, stamp it, log it, and send an email to everyone. I want the engineer in charge to do all that (take the person by the shoulders in a professional manner); plus, I want him to walk over to the field person in charge and let him know that it just came in. I want to make sure that you all understand that relying on emails on this type of fast-track job isn't sufficient. If this was the case, the owner wouldn't need us. We have to have actual conversations with one another if we are going to properly coordinate this job and be successful!

Please don't leave out the most important person while considering the efficacy of your organizational structure. Most projects directors, project managers, superintendents and general foremen grossly underestimate the impact that their frequent absences (due to conflicting job demands) have on their team's performance. For example, the project director for the Seattle football stadium spent most of his work hours in meetings with a plethora of ownership groups. But since he was the person who had written the schedule and held the overall vision for moving the job forward, every time he was out of the trailer (which was roughly 85% of the time), the team floundered. It was not until he found a way to supplant his presence at owner meetings (via the general manager's increased involvement) and make himself more available that his staff was able to tap into his thinking and gain sustainable workflow momentum.

When constructing the organizational chart, it should clearly reflect how the communication pathways *actually* work. It should indicate who reports to whom, provide a brief description of each person's job duties, and the durations expected for each person on the job. In an ideal world, your project administrator or receptionist should be able to glance at the organizational chart and flawlessly route incoming calls. To this end, it's often a good idea to have the project administrator or receptionist fill in the details of the organizational chart. He or she can interview each person as to his or her job duties and create thumbnail descriptions (which he or she reviews with you prior to publication). He or she usually enjoys the assignment and, in so doing, acquires a thorough working knowledge of each person's role and responsibilities. Table 6.1 shows an example of what to include.

TABLE 6.1

Structure

John Brown (ENG)	
On job	**10 months**
Foundations	Concrete
Rebar P/T	Excavation
Shoring	Embeds
Slab edges	

Your organizational chart should be updated on a regular basis, reflecting any additions to the team or changes in role assignments. In this way, it becomes a living tool that keeps discussions about team process and flow in the forefront of your staff's minds. It also keeps everyone's heads in the game about possible changes in communication patterns that may impact their work and what adjustments they may need to make to ensure that everyone is kept in the informational loop. (Later on, we'll discuss how to use your organizational chart to diagnose team problems.)

ROLES AND RESPONSIBILITIES

In 2004, due to a tumult of injuries, the New England Patriots named twenty-two different players as starters on defense, yet they still won the Super Bowl. How did they pull this off? First, instilled by their head coach Bill Bellichick, they acquired an attitude that injuries were simply not a valid excuse for not winning football games. And second, everyone, including the bench players, had a thorough working knowledge of the game plan for each opponent and had a clear understanding of their own role in the context of this plan. The mindset was simple: lacking the physical attributes of a first stringer had nothing at all to do with acquiring a thorough working knowledge of the plan and being able to execute properly if called upon. To ensure this, the coaching staff required each player to pass a written test about the game plan with 90% accuracy. This begs the question: Why should we expect anything less from the people on our job sites? Everyone should know their roles and responsibilities as they relate to the overall job plan—no exceptions, no excuses. Some managers are resistant to the idea of specifically defining roles for fear that some, if required, will not go beyond what is spelled out. But personally, I've rarely found this fear to be grounded in reality. To the contrary, choosing to leave roles and responsibilities undefined or loosy-goosy is a recipe for waste and lost productivity. It is clarity, not confusion, that breeds efficiency and high productivity!

WORK PLANS

Every person on the management team should clearly delineate his or her work plans for his or her area of responsibility and present this plan to the entire team:

> Hi, my name is Mike Stone. I'm in charge of procurement. This is my plan for buying out this job…. Here are my first order of magnitude priorities and why…. Here are my second…. I will be primarily interfacing with the PM and each of the lead engineers as well as the project superintendent. In order to execute this plan, this is what I will need from each of you …, and here is what you can count on from me. Do you have any questions? Does anyone need me to go into any further detail about the plan?

Work plans should be reviewed in stages, so that in subsequent staff meetings, each manager can report on whether or not they are on track.

In terms of Lean thinking, I strongly suggest that you extend the work plan concept throughout the entire project, both internally and externally. Since a construction team's success is predicated upon each person's ability to successfully deliver on the commitments that he or she has made, each person on the team should have a thorough understanding of the work plan developed by the manager within his or her discipline, highlighted by the identification of specific weekly deliverables. As a manager, you should gauge the reliability of these work plans as measured by percent of plan complete (PPC). This next point is crucial: when a deliverable is not completed by the deadline promised, it is time to engage in the five why's (please see Chapter 13). It is important to identify the root cause as to why the plan was not completed, *not* as an exercise in assigning blame, but to discover the objective reasons why it was not completed. Once you have fully ascertained the reasons why the failure occurred (usually an unforeseen obstacle), then you can help the team make the necessary course corrections to help them get back on track.

AWARENESS OF HOW TEAMMATES IMPACT ONE ANOTHER

This point bears underscoring as it is often neglected in the establishment of roles and responsibilities. Everyone needs to be cognizant of the fact that they *do not* work in a vacuum, that what they do—or don't do—will have a definite impact on those around them. A responsibility matrix should be issued so everyone on the team can gain a thorough understanding of what their teammates are doing and how they are impacted when logs are not maintained and procedures are not properly followed. When we talk about teammates being dependent upon each other, this is precisely what is meant. People depend on others doing their portion of the work in both a correct and timely manner so they can do theirs. The PM, PS, and GF should take every opportunity to verbally reinforce this reality at every turn. Should you notice that a protocol has been breached, call an impromptu meeting:

> Listen up people, this is why replacing current documents to their proper spot and noting their return is so vital. Carlos just wasted an hour tracking down a miscellaneous metals drawing that has apparently been sitting in the back of the conference room for the past three days! As a result, he was late getting the updated information to his subcontractor. Let me clarify something; it is not the sole responsibility of the document control coordinator to keep all of our plans in order. It's on all of us! This is not how we run an efficient and productive ship. What can seem like a little thing isn't so little when it wastes someone else's time.

Another way of building this awareness is to create what is often referred to as a war room. It is a room specifically designated for posting the organizational chart, the master schedule, schedule updates, and all pertinent work plans. At a glance anyone on the project should be able to walk into the room and briefly ascertain the project's progress and who is responsible for doing what without having to search through multiple binders in various offices scattered throughout the trailer. If updated regularly, the war room will increase the accuracy of coordination efforts by enhancing the team's ability to consistently go to the right people with information or questions, rather than wasting time by engaging the wrong ones.

Know Your Audience

It's vital that you gain a working knowledge of your staff's various skill sets ASAP—not just in terms of their technical prowess, but also for how best to interact with them. Believe it or not, this is the most efficient way to determine how best to use yourself as a leader. Not every team is composed of the same types of people; each requires a slightly different focus and approach. The following case example illustrates this point. A senior project manager with twenty-five years of experience with his company found himself in a quandary. He inherited a team that was long in the tooth in construction experience, but short in years of experience with his company, and they suddenly were balking at his leadership style—frequently accusing him of being a micromanager. But from the PM's perspective, they were making simple mistakes, and quite a few of them, so it was easy for him to dismiss their complaints regarding his leadership. In his mind, his behaviors were justified based on their "screw ups," and he was beginning to have doubts about just how experienced his teammates actually were. The team was at a seeming stalemate. The more the PM attempted to micromanage them, the more they resisted his efforts to manage them, which resulted in even more mistakes and, consequently, even more micromanagement by the PM—a classic recursive cycle (i.e., a negative, self-reinforcing cycle where one undesirable behavior leads to another). To break this type of cycle, it is important to look for the objective reasons for the difficulties. Just a little bit of digging into the PM's history quickly revealed the root cause of the problem. For most of his career, the PM managed people who were straight out of college—and did a terrific job leading them. Given how overwhelmed they often felt, most of these "green as grass" underlings welcomed his highly directive approach and didn't mind his micromanaging style at all; in fact, most viewed him as an asset to their careers. But the veterans on this new team held an entirely different view. They knew what they were doing and were eager to demonstrate this to their new company. What they lacked was the ability to translate their knowledge and experience into the language of their new company. So how did this situation become so misaligned? When they did things incorrectly, instead of honoring their experience and helping them to translate what they knew into the particular vernacular of their new company, the PM would launch into long lectures about what they should have done differently, and what he would have done back when he was in their shoes. Not surprisingly, these lectures didn't improve performance. In fact, they made it worse. People were so angered by the PM's lack of acknowledgment of their expertise that they simply avoided him, which only served to compound the errors that they were already making. This, in turn, lead to more micro-management by the PM. To the PM's credit, when he was made aware of how his approach was negatively impacting the team, he changed his attitude and behavior toward them. Instead of assuming that a lack of basic knowledge was the root cause or their execution errors, as he had been, he instead asked them what they were attempting to accomplish. He then helped them to connect the dots within the framework of the company's policies and procedures. As a result, the team dynamic changed almost over night. Instead of avoiding him, the staff began to seek out the PM's feedback and advice. More importantly, the quality of their work improved at a rapid rate.

RESPECT FOR CHAIN OF COMMAND

Believe it or not, this isn't just a respect issue—it's also a productivity issue. Bypassing the chain of command may seem expedient, but in the long run, it causes untold negative disruptions. A PX who was instrumental in selling a $375 million hospital job in a remote area unwittingly demonstrated just how disruptive breaks in chain of command can be. A PX had an excellent relationship with the owner, and since he wrote the master schedule, his expertise on the project was unassailable. But given the location and the fact that he carried multiple assignments, meant he spent at most only one or two days physically on site. Not being one who shied away from making decisions, if he observed something that made him uncomfortable, rather than voicing it to the PM, he'd immediately redirect a junior staff person to correct the issue to the manner of his liking. Now, you might well ask, what's the problem with this? Wasn't the PX simply being an engaged and expedient problem solver? Yes, in some ways he was. And he certainly did have the project's best interest at heart. Unfortunately, this method of problem solving often creates more problems than it actually solves. Since the PX rarely looped back to discuss his redirections, when the PM checked on the progress of an activity that he had assigned, he'd discover, to his dismay, that the underling had dropped his assignment in deference to the PX—thus throwing his own game plan completely askew. To complicate matters further, the PX and PM were both twenty-five-year veterans of the company and had a great deal of respect for one another's reputations, so neither felt comfortable letting the other know that he didn't appreciate what the other was doing. Instead, they simply went around each other's backs, taking turns countermanning various orders, while the resentments they held inside for each other grew. Now, think about how this unspoken conflict impacted the production line. Instead of receiving a clear and unified direction from the management team, the staff was often confused as to whose direction they should follow—and what the overall plan truly was—because it always appeared to be in flux. Productivity slowed to a crawl as people asked themselves, "Who is the real boss of the project?" and "Who does my performance evaluation?" (i.e. "who should I actually listen to?") Rather than focusing on execution, they sought out strategies to best protect their own bacon.

The good news was that once informed that their staff felt like kids whose parents were going through a divorce (i.e. trying to figure out which parent they wanted to piss off least), the PX and PM were duly appalled. They immediately arranged to meet once a week, along with the rest of the management team, to discuss "hot issues" and to make sure that they were all giving a unified game plan to the team. They also agreed to make it their personal responsibility to adhere to and honor the chain of command. To their credit, they stuck to this plan for the remaining two years of the project—with excellent results. Fortunately, this problem was arrested early or before it had become overly detrimental. Can you imagine how costly this would have proved to have been if this had been allowed to flourish throughout the life of the project?

In terms of chain of command disruptions, it is also important to be mindful of those created by the owner. A classic example was a joint venture between a Dutch and U.S. company to build a fast-track pharmaceutical plant in the Netherlands for an American owner. Just four months into the project, the ownership group had become extremely frustrated with this joint general contractor (GC) effort, accusing them of not executing the job in accordance to a cohesive plan. Upon examination, there certainly were some critical team execution issues. But what also became clear was just how much the owner was unwittingly contributing to this problem.

The job was set up to have a single point of contact between the owner and the GC—the owner's representative. Technically speaking, all communication between these two entities was supposed to flow through one individual. But when I asked the GC management team how this single point of contact system was working, they all laughed to the point of tears. They quickly informed me that everyone on the eight-person ownership group called "anyone they damned well pleased" on a regular basis. A typical day for the GC's project director (PD) looked something like this: Two sips into his morning coffee he would get a call from an irate owner. Ten minutes earlier, the owner had spoken to a young engineer looking for a specific piece of information and was greatly displeased by the answer he received. Caught off guard, the PD would then have to spend the next hour either researching the owner's question or calming them down and assuring them that they had been given incorrect information. He would then have to spend an additional fifteen minutes fending off accusations that either he was lying or the subordinate in question was incompetent. In the meantime, he would see his voicemail messaging system light up like a Christmas tree as additional owners cued up to register their displeasure. This process replicated itself like a virus until sundown, when the owners finally left for the day and the PD could get some actual work done. I recall thinking at the time that it was fortunate that Holland has such strict bans on handgun ownership, because the PD was more than ready to start taking hostages. You could certainly understand his frustrations. He was being criticized for the very thing that the owner was inadvertently preventing him from doing. Any hope that the project director had of formulating and executing a plan with his team was blown out of the water by 8:10. And this happened every single day, five days a week. This was exacerbated by the fact that the engineer of record was back in the United States, and given the six-hour time difference, everyone was behind the informational eight ball.

As uncomfortable as it can be to do, limiting owner disruptions by truncating unnecessary meetings, adhering to specific points of contact, or redirecting your staff to deflect all calls that should come to you is the best way to prevent unwanted disruptions of this type. And this isn't being selfish or uncaring toward the owner. Just the opposite. By minimizing disruptions, you will be increasing productivity and eliminating a key source of waste. In essence, you are protecting the owners from themselves, which, believe it or not, is an ethical "ought" of your job. But should you find it necessary to take such actions, please don't omit this crucial step: educate the owner as to why you are limiting his or her contact with the project and what you are trying to prevent in Lean terms. You are doing it because it is in his or her best interest not to disrupt the flow of the job!

POLICIES AND PROCEDURES

One of the biggest Lean killers in the industry is the preponderance of paperwork that sucks untold man-hours from every office and job site in order to mitigate risk. It is the single biggest source of waste and lost productivity on a job, a situation that only seems to get worse each year. It is ironic that our very efforts to mitigate risks are the ones that lead to our greatest source of waste, fueled by the ever-growing legal industry. In California alone, from 1993 to 2003, the number of construction law attorneys in the state rose from four thousand to ten thousand—a truly staggering number. A few senior managers in California were so disturbed by the trend and curious about its impact that they actually tracked their hours for an entire month. What they discovered shocked them. In any given week, they spent 85% of their time satisfying internal and external reporting responsibilities that were not at all value added (i.e., contributed nothing to the actual building process.) One can only shudder at the thought of so many woodland creatures rendered homeless just so construction companies can build paper fortresses to protect themselves against potential lawsuits.

Having said this, not all paperwork is unnecessary or wasteful.

Many well-run companies spend a great deal of time, effort, and money establishing solid policies and procedures that do advance the building process. If developed and executed correctly, policies and procedures create a responsibility chain that, similar to checklists in manufacturing, ensures a smooth flow of production all along the line. In such companies, top managers are able to examine the paper trail at any given point in the chain and determine exactly whether the job is progressing as projected. Despite these efforts, there are still some managers who hold the belief that "no paper is good paper." I have known managers who, when confronted about their rebellious lack of adherence to policy and procedure, insist that they can keep all of the important agreements and promises in their heads, and in fact view this as a mark of distinction—even going so far as to belittle subordinates who can't (or won't) do the same. If you are one of these proud types, I have just one question for you: What happens to all these lovely agreements and promises that you have made throughout the course of the job if, God forbid, you were to be run over by a bus at 2:00 p.m. this afternoon? Wouldn't they go straight into the grave along with your lifeless corpse?

That is the real point of well-crafted policies and procedures. They don't just create a useless paper trail. They ensure that everyone along the assembly line has the correct information to do their jobs effectively. It is yet another way to minimize disruptions and keep the line moving!

Here's another way to think about policies and procedures that that may be less apparent. No company is composed entirely of A+ players. In fact, following the laws of normal distributions, the larger the company, the more likely it is that it will have a fairly sizable distribution of B and C players as well. Effective procedures are a means for A+ players to pass down their knowledge and expertise to the rest of the company. In a very real sense, through these procedures, they are able to convey to those who are less experienced (or perhaps, less motivated) "how to think" about the job that they are doing. By following policies and procedures precisely, B and C players can execute nearly as well as A players, even though they lack their in-depth knowledge and experience. (Later on, we'll talk about how to use policies and procedures as an assessment device to diagnose team problems.)

Tools to Do the Job

It is one thing to have proper policies and procedures in place, and quite another to know how to utilize them. If confusion exists about how to properly use your operating systems (Prolog, Suretrack, Accubid, etc.) or any other standard practice, it is as if the tool does not exist at all. Have a training plan at the ready for those who are new to the industry or to your company. This is particularly true if you have systems that are idiosyncratic to your company. This includes whatever unique acronyms your company might use. Relying on "catch if catch can" methods for passing on this knowledge will almost certainly result in significant losses in productivity until people are able to get up to speed. One company has a unique way of handling this type of issue from the start. Whether the recruit is fresh out of school or is a grizzled veteran, he or she is not permitted on site until he or she has gone through an extensive two-week "boot camp" that consists of a formal orientation program and rigorous training on all company policies and procedures, which employees are actively encouraged to take more than once. They have found over the years that this up-front expenditure provides a great return on their investment by helping new employees become fully functional and acclimatized in the earliest stages of their employment. It also has the side benefit of weeding out the truly clueless ASAP.

In terms of "hard tools," anything, within reason, that increases productivity and prevents team missteps should be considered value added—even if it is a bit pricy. Walkie-talkies and radios are a great way for people to close the informational loop in real time, while emails, which are convenient and relatively inexpensive, have very limited value. Emails are great for tasks that require simple one-way communication (i.e., announcing a meeting time or when a conversation needs to be documented and confirmed). But all too often, people supplant actual interactions with emails and, instead of picking up the phone and engaging in healthy debate, issue an edict that the other person invariably interprets the wrong way. If you truly want your process to flow, insist that for all important correspondence, particularly those that require a decision to be rendered, your people get off of their posteriors and have an actual face-to-face (or phone) conversation.

At the job site, consider biting the bullet and investing a fair amount of your fixed operating budget on your plan room. There is no greater way to observe waste than by watching people search for documents that they can't find—or far worse, sending jobs out to bid based on obsolete drawings. It's one of the first meaningful steps you can take to ensure high productivity. Projects that have solid document control procedures in place, and solid people to manage them, are simply more productive.

In this same vein, demand clean, orderly, and well-maintained work areas. Your trailer needn't resemble an army barracks, but managing by piles is a needless waste of time and energy, unless, of course, you actually want to create a bad impression in the minds of your owners, architects, and subcontractors who visit your trailor.

FLOW

I learned the concept of flow from Gus Sestrap, a project executive with Turner Construction. Put aside the fact that Gus writes his schedules by visualizing the completed project in his head, then deconstructing it backwards—which is just plain scary brilliant—and consider his resume. His teams consistently deliver projects (including $350 million stadium and hospitals jobs) ahead of schedule. Gus's philosophy is freakishly simple. He insists that a well-run job is *not* one where subcontractors are pushed to the breaking point, but instead, it is about *flow*, that is, establishing a logical, predictable, and doable rhythm that is able to be sustained throughout the life of the project. With subcontractor input and buy-in, he constructs a schedule that can be used as a reliable tool for material procurement and manning the job, and gives the owner a clear picture of how the job will unfold. Inherent to his philosophy is the establishment of a buddy system between field and engineering functions. Each activity has a corresponding field and engineering person assigned to it to ensure that the project consistently is built on paper *before* any actual construction activities begin. (The equivalent for subcontractors is to ensure that project managers are walking the field in concert with their general foremen on a regular basis.) When this system is working properly, RFIs are fully vetted *before* they have a chance to negatively impact the job.

Unfortunately, I've witnessed the opposite philosophy play out on far too many jobs—with disastrous results. For instance, on a $400 million public hospital OSHPD project in San Francisco, a GC decided to put a hard-nosed, hard-driving, independent-minded superintendent in charge— someone who did not to believe in working closely with his engineering counterparts. For the uninitiated, OSHPD is the California agency that, by law, *must* review and approve all structural elements and changes—for every single hospital built in the state. It is specifically charged with ensuring the seismic integrity of every project that comes under its purview. To be fair, this agency has its place, but it is chronically undermanned and underfunded, so it is not unusual for a project of this type to come to a screeching halt as revised drawings and specifications pile up on the agency's collective desks for review. In such a scenario, a well-meaning but hard-driving superintendent who is constantly pushing subs to get work in place as fast as they can, but isn't fully engaged with the engineering side of the house, does nothing but run subcontractors straight into costly and unproductive brick walls. Relentless pushing creates workflow patterns filled with stops and starts and that are fraught with waste and failed work plans. Given that manpower is the greatest risk that any subcontractor carries, you can imagine how difficult it was for this GC to get these same subcontractors to bid on future phases of this project, at a reasonable price.

PULL VERSUS PUSH

When we push we miss golden opportunities to engage in Lean thinking. It is much more effective to pull than push. What do I mean by this? When we push people to go as fast as they can, we are, in essence, driving them toward a result regardless of the obstacles they may encounter. What's wrong with this approach you ask? Plenty. The people we push may eventually get to the result we want, but we may also sustain a great deal of collateral damage in the form of incurred waste as a by-product. When we chose to pull people instead of push, we are actually asking them to consider and weigh the obstacles in front of them and formulate a plan that allows them to move past them with minimal waste. Here is how it works. Between milestones, gather your subcontractors and ask them to brainstorm in terms of what they will need, and what will need to happen, in order for them to hit the next milestone. Instead of purely driving to a result, this type of thinking invites key players to engage in preplanning and strategizing. It is far more useful for identifying and clearing obstacles, and as a result, the players will be far more likely to achieve the result you are seeking. It is certainly far superior to bellowing, "I don't care how you get there, just get it done!" By looking to the right (scheduled milestone) then shifting to the left (what is needed to get there), subcontractors can make the necessary midcourse corrections to ensure that the targeted milestone does not fail. By actively engaging the key trades and treating them like valued customers, you will be inviting them to be part of a solution, as opposed to waiting for their inevitable failure and treating them like adversaries. In so doing, you will also be able to prevent potential claims.

THE SCHEDULE

Besides the fact that it is a requirement, the schedule is your primary tool for helping others to visualize flow. In terms of our assembly line analogy, the schedule determines the speed at which this giant conveyor belt will move, as well as conveying the vital activities that are required along the way to feed the line.

First and foremost, the schedule should be realistic. For instance, if you have an OSHPD job as described above, float time for potential delays needs to be built in from the start, particularly for areas that you anticipate significant redesign. If not, there will be a bust in your schedule before you even break ground. An effective schedule also anticipates the time required to allow critical engineering and design functions to get out ahead of construction (i.e., time needed to process submittals, change orders, RFIs, and engage in value engineering). You should also take into account the decision-making capabilities of the owner. Private jobs manned with experienced owners usually have greater and faster decision-making capacity. For them, time equals money, so they tend to err on the side of expediency. The public sector is another animal entirely. Your counterparts in this world are well aware that they can't get fired for saying no, but they can get fired for saying yes. So you can anticipate long lag times whenever key decisions need to be made. After all, it takes time and effort to spread the risk of potential negative repercussions onto your fellow coworkers!

Make sure to take the time to fully educate your staff regarding the schedule. The best schedule in the world is useless if the people on the project don't fully understand it or fail to grasp its implications. You might as well not have one at all. Review it thoroughly and via planned updates; take every opportunity to bolster people's comprehension. I guarantee that your up-front efforts will pay big dividends in the long run.

This last point is critical. Make sure to actually publish the schedule. The correlation of troubled jobs paired with the failure to produce (or update) the schedule is staggeringly high.

Flow and the Individual

Let's shift our focus back inside the trailer. When considering flow, it is also important to evaluate each of your teammates. In his ground-breaking book entitled *Finding Flow: The Psychology of Engagement with Everyday Life*, Mihaly Csikszentmihalyi describes flow as the internal state of optimal productivity. It is the exact opposite of the Peter principle, where people rise to their level of incompetence, thus killing all forms of Lean thinking. It is when what is asked of people matches what they actually know how to do. Conversely, if people are asked to do something they don't know how to do (state of feeling overwhelmed), or are asked to do something far below their skill level (state of boredom), they are far less productive. What does this mean to you as a leader? Simply put, if you have a lot of people who are not in an internal state of flow, the overall workflow of the project will be compromised. To counteract this, you need to meaningfully assess each person's skill level as measured against the unique demands of the job, and based on your findings, either provide necessary training, shift people into positions that better suit their skill sets, or manage upwards to acquire those who possess a better skill fit. You may need to make a little noise with your superiors in order to acquire what you need. Remember: It is the well-informed squeaky wheel—the one that is prepared and makes a good case for what it needs—that often gets the grease!

FEEDBACK AND POSITIONING

Often, due to circumstances beyond their initial control, leaders find themselves behind the eight ball when jobs start. As a result, they fall into a reactive posture, playing a seemingly endless game of catch-up. This usually happens when they are named late to the job or the project breaks much quicker than anticipated. Leaders in such situations often give a cursory nod to objectives and milestones in staff meetings, but more often than not, they find themselves locked away in their offices, cranking out mountains of deliverables while barking out various demands that make little sense to the people they are managing. As a result, their stance toward their own team is often one where they are standing behind their people, pushing them toward an objective that, from their staff's perspective, is fuzzy at best. And when the leader does stop long enough to give feedback to his or her staff, it is usually to inform them that they missed the mark by a wide margin. While pushing does impel people to work harder, since they lack the vision for the overall product and the context for many of the activities that they are engaged in, pushing also tends to produce a great deal of waste in the form of improperly performed tasks.

I'd like to suggest a more productive alternative, and demonstrate it by going back to our kitchen analogy. Remember, the goal for any restaurant is to have a successful diner service. This means that each successfully plated meal is a subassembly, and each table in the restaurant represents a milestone toward the goal of a successfully completed dinner service. Pulling all of this off requires many hands. To start, menus have to be developed and printed, ingredients need to be purchased, the wait staff needs to know the menu and the specials, and the various chefs have to prepare their individual workstations. Further, the bus boys have to set, clear, clean, and reset the tables numerous times throughout the night. The key for accomplishing all of this is for everyone to do their part to communicate what they need in accordance to the overall service goal. But coordinating all of this requires a bit more than simply communicating.

For those of you who watch *Hell's Kitchen*, do you notice where head chef Gordon Ramsey positions himself? Usually, he's not in the kitchen, but at the end of the pass line where all the fully plated dishes arrive. Why does a head chef position himself there? It's actually quite purposeful. Since the head chef is the person who constructed the menu and holds a clear vision of what the finished product should look like, standing at the end of the line serves as a final quality control point. Incorrectly plated meals (polenta instead of the intended risotto as a side) or over- or under-cooked meats can be detected and corrected *before* the product goes out to the customer. (By the way, in any process, you always want to catch your mistakes *before* the customer does.)

But there is an even more important reason for standing at the end of the line. Because the head chef not only knows what the finished product should look like, but also what is required to produce it by positioning himself at the end of the line, he or she can observe the team process as it works its way toward the final result. He or she can literally track execution, coordination, and timing—and provide pertinent, real-time feedback to ensure a successful outcome. Hence, the head chef can urge the meat person to speed up, ask the vegetable person to slow down, or ask the potato chef to put aside what he or she is doing and lend a hand to another chef. The head chef can also coordinate with the waiters to push the specials if there is an inventory surplus, or to take items off the menu if there is an unexpected run on them. While you may find Gordon Ramsey's interpersonal tactics a tad on the abusive side, do take special notice of what he harps on his teams to do the most: "Talk to each other!" He knows that without the additional element of communication, the cooks in the kitchen have no shot at coordinating their actions and completing a successful dinner service.

Let's look at this same scenario from a push perspective. If all the head chef did was stand in the back of the kitchen and bark at each chef to work harder, they would likely heed his call and crank out an ungodly amount of product, but this would be no guarantee of success. The far greater likelihood is that they would produce enormous piles of food in their particular area of responsibility that would become spoiled, thus generating a tremendous amount of unused inventory (waste). This unused inventory would eventually be thrown into the dumpster, with only a small percentage of what was produced winding up as successfully plated meals! Figure 6.1 shows how all this would look graphically.

Though we often fail to recognize it, the exact same rules apply in construction. What is the point of an engineer cranking out one PCO after another, or the superintendent pushing subcontractors to work as fast as they can, if, because they have failed to coordinate their actions, mountains of paperwork are produced that are never looked at, or the work performed is so out of sequence that it impedes other trades?

Instead, try standing just beyond the upcoming milestone, and pull your team toward the result. As you did with your subcontractors, use staff meetings to identify what needs to happen, share your concerns for readiness, and make the necessary preparations as required. Feedback given in this manner is in real time, and is largely proactive rather than reactive. Also, make sure to leave room for your staff to express their concerns and what they will need by encouraging their forthright honesty. By obtaining real-time status reports on completed work plans, you'll be able to gauge if course corrections are required, or where you may need to shift (human) resources to ensure that a milestone does not fail.

There is another advantage to delivering feedback in this manner. It gives your team a context by which to judge their own actions. Because they are working on real issues, the feedback provided won't be as readily tossed aside as so theoretical posturing. Think about how easily the constant harangue of "Okay folks, it's important to keep up with our RFIs" can be dismissed. But if instead what is heard is "Okay, Sarah,

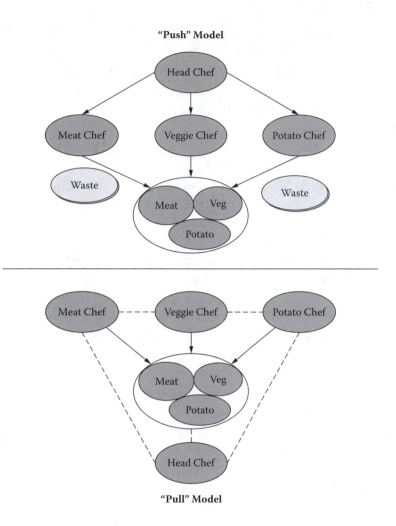

FIGURE 6.1
Push model.

where are we with RFI 216? We need to have an answer by Wednesday or Juan at EFG plumbing is going to be impacted! What are you going to need in order to get this RFI processed on time?" not only will you stir up a needed sense of urgency, but because she now has the proper context, Sarah is far more likely to learn the requirements of her job—both now and for the future. And because feedback is immediate and in context, it allows her to reflect and take action *before* disaster strikes. There is nothing intuitive about writing a PCO or processing an RFI. People have to learn what quality RFIs and PCOs look like, and they learn these tasks best if they have a context to place it in.

This points to another side benefit of the pull model: It also allows you to give real-time *positive* feedback and praise for successful planning and execution—a tool that is invaluable for establishing a Lean culture.

Positioning yourself in this manner also allows you to manage information and the plethora of decisions that need to be made much more effectively. There is usually such a flood of both that most teams gain a decided advantage when the person at the top is able to filter the chaff from the wheat and keep what is truly important in the forefront of everyone's mind.

Please do take the time to give thorough feedback offline and see it as the valuable tool it can be—particularly for your inexperienced personnel. If they know what they are doing right, and what they are doing wrong, then they will be far more likely to hit the desired target. This doesn't have to entail a lengthy or elaborate discussion; it just has to include the following:

- How their performance is measuring up against established expectations
- What they will need to do to improve performance
- What they are doing right, and in what ways they are meeting or exceeding expectations

Also, include global feedback on the following:

- How we are performing per the schedule
- How we are performing per the budget
- What the owner is pleased with
- What the owner is upset about

This will help to broaden your team's awareness of the context of their actions.

While on the subject of feedback, don't forget to build in the most important feedback process of all—the loop that extends from your team to you! If you've extended the proper invitation with the proper attitude, this will be a breeze. Everyone on the job should be encouraged to be active engagers versus passive witnesses. Therefore,

- If they detect obstacles in the process that they can't solve themselves, they need to speak up.
- If they have been inadvertently given conflicting directions from various managers, they need to point this out.
- If they are suddenly performing multiple roles and this impedes their productivity, they need to raise their hands.
- If the chain of command has been broken, they need to raise this awareness.
- If unnecessary meetings are intruding on their ability to execute their roles, they need to ask for your help.

But of course, the usefulness of a team's feedback rests entirely on the leader's ability to see it as a vital contribution versus a threat to their standing or authority. If the leader's ethics and motives are pure, he or she should have no problem making this distinction.

One important caveat: If anyone on your management team has the audacity to say, "Yeah, I saw that coming—I knew that was going to be a problem," and didn't speak up beforehand or do anything to prevent the problem from occurring in the first place, you have my permission (provided that it is done professionally) to flog them within an inch of their lives!

Attending to the basics is the best way of preventing the type of problems that plague most projects. Unfortunately, most managers feel as if they don't have the time to set things up right. Let's interrupt that thinking now. If you want to think and be Lean, you don't have the time *not* to set up the job correctly!

7

Execution and Overarching Philosophies

As an extension of the basics, the execution and overarching philosophies further establish the ground rules that everyone on the team needs to play by. Do you know those complaints you hear about people not understanding the big picture? This is what they are actually talking about. They want to know *how* they should be approaching their work per the contract (the execution philosophy), and what they should keep in the forefront of their minds based on the specific promises made to the owner (the overarching philosophy). Though these philosophies may conflict, it is essential that you help your staff synthesize them. Even if your contract isn't integrated project delivery (IPD), opportunities for Lean thinking abound provided that both philosophies are explicit and the focus of everyone's attention.

OVERARCHING PHILOSOPHIES

Why was this job awarded to your company in the first place? I'm guessing that it had something to do with how well your top management team listened to your client's needs and concerns, and what they believed your company could do to address them. When not bound by price alone, owners select general contractors (GCs) and contractors because of their perceived ability to handle a variety of specific variables, such as

- Unique technical requirements (i.e., high-tech tool installs, clean room experience, biotech experience, OSHPD hospital expertise, etc.).
- Specific efficiencies (i.e., transparent accounting methods, project management tools [i.e., Prolog], prefabrication capacity, Lean construction methods, etc.).

- Sustainability factors (LEADS certification, etc.).
- Staff! Staff! Staff! Many clients know that a company can do the job, but are they willing to lock in the people they proposed on the project?

The owners' needs can also include concerns that they are reluctant to overtly express for fear of appearing too vulnerable, such as

- Please help us manage the things that we have no idea how to manage (we don't really understand construction).
- Please help us keep the costs down (we don't know the market and actual costs).
- Please help us identify the needs that we didn't even know we had (did the initial design take our growing capacity or limited budget into account?).
- Please help us get the best possible deals (again, we don't know the market).
- Please do all of the above throughout the life of the project (don't abandon us at the end of the job!).

Everyone on your team should know what is important to the owner, and on what promises the contract was sold, because these will establish the baseline expectations by which the team will subsequently be judged by the owner. So, if your crack business development team promised state of the art 3D BIM, or prefab capabilities second to none, there should be no subsequent questioning of their importance, and certainly no failures in delivering them. After all, this is a matter of trust. To fully understand this from the owner's perspective, let's look at an analogy. Let's say you are remodeling your kitchen. What would it mean to you if you said to the person that you hired that above all else, what you wanted to see first was a variety of cabinet facings, and they promptly showed up the next day, at 8:00 a.m., with twenty cabinet samples in hand? The immediate impression you would have is that the person not only heard you, but that you could begin to take him at his word. You could relax—at least a little. Your anxiety-driven need to scrutinize the person would start to go down. Now imagine instead if this same person arrived two days late, armed with nothing but excuses and countertop samples. Your first and enduring impression would be that you'd made a big mistake in hiring him and that you will now have to watch him like a hawk throughout the life of the job.

Rightly or wrongly, I find that judgments about a GC's or subcontractor's integrity are usually determined within the first few weeks of the job. Joe Takash, president of Victory Consulting, is fond of saying, "Words are like toothpaste; they are very hard to get back in the tube once they are out." Well, the same can be said of first impressions. We are judged not by our "first date" behaviors that we exhibited during the presentation, but by *how* we delivered on our promises at the very *beginning* of the job. Our true intentions are believed to be revealed by our initial actions. And once the impression is set, it is very difficult to alter. Circling back to our discussion about biases, failing to fulfill a promise early on automatically cues up all of the owner's predisposed bad thoughts about GCs and subcontractors, and will register loudly as a hit on their danger-detecting radar screens. Once registered, the owners will continue to scan for similar behaviors and will feel increasingly justified scrutinizing you and your team's subsequent actions. In Lean terms, the impact caused by such missteps can be catastrophic. Once mistrust is cued, a number of unwanted behaviors, such as posturing (versus an honest exchange of ideas), increased scrutiny (having to justify every action taken or not taken), and delayed decisions (preoccupation with ferreting out dishonest numbers), will be exhibited, and all of these will lead to lost productivity and increased waste.

By the way, this notion of first impressions goes both ways. If you produce a mock-up on the ground level with easy access, the owner needs to understand that the time and cost of doing this work 100 feet in the air, off of a scaffold, in poor weather conditions will be different than if it is undertaken in ideal conditions. It's important to spend time with the owner to explain that situations like these aren't "hits" on their radar screens, but legitimate conditions.

EXECUTION PHILOSOPHIES

Execution philosophies are largely dictated by the type of contract that has been signed. Each delivery system carries with it a set of unspoken expectations about how the work is to be carried out. Because they have become second nature to them, most experienced leaders take for granted the amount of understanding that their subordinates have of contract terms. The reality is that the grasp isn't nearly as strong as they believe it to be—particularly around the subtleties of each delivery system. And it is not just inexperienced staff members who struggle. People with a great deal of construction experience, but who have worked with companies that offered limited contracting modalities, can struggle mightily—and are least likely to let you know it for fear of appearing foolish or unqualified. While what follows is surely old hat for most of you, I hope these scripts serve as a reminder of the things to cover as you bring your staff up to speed.

Traditional GC role: As the GC, it is our job to be the driver, that is, to be out ahead of the project in terms of budget, schedule, quality, safety, and buy-out. To be effective, we need to know our scopes, get the job out to bid, review shop drawings, and get the building built on paper as quickly as possible. The team's particular focus needs to be on identifying hiccups like long lead items or funky complications in the drawings that need early clarification. It is also our task to create, with subcontractor buy-in, a master schedule that establishes a dependable workflow and alerts the owner to specific areas when critical decisions will need to be made. As the GC, it is incumbent upon us to make sure that the owner has all the information required to make informed decisions—well before the issue has reached a critical stage. In this role, "letting the job come to us" is a recipe for disaster. Everyone's focus should be on studying the drawings, reading their contracts, and working hard to get the job set up right from the very first day. (Any lack of urgency at this point should be adjusted quickly and firmly.) "Okay team. I'm sure each of you is diligently studying your plans and specs, so what I'm about to ask you shouldn't be that difficult. By next week's staff meeting, I want each of you to prepare a ten-minute presentation about your area of responsibility, your game plan for executing your work, and most importantly, what you will need from others on the team in order to effectively execute your work plan." (Let them know that they will be asked specifics, so bluffing will not be an option. Nothing raises a sense of urgency like the specter of doing a little public speaking in front of one's peers. The other purpose of doing this is to once again drive home the awareness of how each person impacts everybody else on the team.) "Okay folks, I think Jennifer has a damn good plan for keeping the submittal log updated. But it all falls apart if we don't continually feed her the information that she needs. And if it does fall apart, then when Dennis tries to implement his work plan in the field, he'll be unclear as to whether or not he's installing the proper materials, so he'll to have to stop dead in his tracks to hunt the information down."

Traditional CM: While often performed by GCs, this modality requires us to adopt a different mindset. Traditionally, in this role, we are not the driver. We are literally an extension of our client. It is our job to administer the project and ensure the delivery of plans, specs, and contracts. But the final direction comes from the client. We are tasked with gathering all the information necessary to allow the overall project team to stay focused and organized, and to help the owner to make wise and timely decisions. This means that our primary job is to obtain all the information that the owner needs in order to make that happen. Our approach needs to be collegial. While we understand the GC's pain, our job is to be a trusted advisor to the owner. This sometimes includes letting the owner know (tactfully) when they are cutting off their own noses to spite their faces should they get into conflict with the GC on any particular issue. But heed this important caveat: if we become overly confrontational, pushy, or directive—particularly with the GC's subcontractors, with whom you should have no direct contact—we'll gum up the entire works, and drive a wedge between all of the entities involved. (This is where some of the old hands, particularly those who have been used to working in a lump-sum environment, truly struggle. Not being the driver just doesn't feel right to them.) Our role in this contract system is to resolve unproductive conflicts—not create them—and most of all, keep the process flow moving at optimal levels.

CM/multiple prime: This is a hybrid CM role. We are responsible for producing a master schedule, keeping it updated, and for keeping the paper processes moving. But the responsibility for coordination, direction, and key administrative functions shifts to the subcontractors who are in privity (direct contract) with the owner. In a very real sense, the subcontractors become mini-GCs of their work. The CM in this delivery system resembles an orchestra conductor, who also happens to own the master schedule (the sheet music, if you will). Unfortunately, most jobs of this type tend to misfire, as people within the various entities unconsciously fall back to familiar roles and begin acting as if they were in a traditional GC model. Conditions (expectations) for this type of job need to be established early, then reinforced, so everyone can adjust to their nontraditional roles. (Your staff need to be explicitly coached to stay within the scopes of their contract. But they need to find a non–passive-aggressive way to express why they are not acting like a traditional GC versus simply choosing to not return phone calls in order to "teach subcontractors a lesson").

GMP (guaranteed maximum price): The great advantage of this delivery method is that it allows us to move money around from pot to pot. The overall focus is to save the owner money. Since the overall price has been negotiated and guaranteed, the owner is less invested in worrying about what pot we pull the money from. In other words, if we need to take some money allocated for drywall (where we are experiencing a savings), we can shift it over to light fixtures (where there was a bust in the estimate). We can do this without the usual hassles you'd experience in a lump-sum environment. The key from a Lean standpoint is that the field and engineering functions need to be completely open and honest about their budgets and the work that has been put in place to date. (I once saw a team struggle because a project superintendent [PS] wasn't used to shifting money around. With the best of intentions, he continually jammed up the subcontractors, when he should have gone to his PM and let her know about the difficulties he was having. For her part, the PM kept accusing the PS of not being a team player—not recognizing that he had never worked in this type of delivery system before.)

It is in this environment where we can practice the lost art of construction. We can use our relationships to help a contractor, while at the same time getting something back in return later on down the road. (One story typifies this type of artistry. A superintendent for a GC heard complaints from subcontractors about their inability to easily access the site. So he scoured the contract, located some money, and authorized funds for a ramp to be built. He didn't have to do this—it wasn't in the GC's contract to provide this—but it ended up paying huge dividends. The subs were much more efficient and productive, and when he needed some change orders to disappear down the road, the subcontractors had no problem accommodating him because of the savings they had realized by working more productively because of the ramp.)

Speaking of change orders, unlike the lump-sum environment, for GCs there is no money to be made from change orders in a GMP—they are included in the overall price. That's why it is so important for us to be looking ahead as much as possible—to prevent change orders from happening. The project makes money by being efficient. Since the price is guaranteed, any savings realized by saving on labor costs or improving efficiency is money in our and the subcontractors' pockets. Thinking Lean within this modality equals profit.

In terms of the owner, the biggest hiccup on this kind of contract centers around what the term *GMP* actually means to them. The price includes what is delineated in the contract documents. It does not include any unforeseen conditions or what the owners or architects decide to add or change later on. We need to clearly underscore "what is in and what is out" at the very outset, or the owners will think that we are pulling a fast one on them when we hit the first unforeseen condition and try to collect money for it.

The other plus of a GMP for the owner (and do make sure to point this out) is that this delivery system allows us to select from a choice pool of preferred, reliable, and professional contractors—rather than being stuck with just any low bidder that comes along. This means (at least theoretically) that we can utilize the relationships we've developed over the years to deliver a smoother, faster, and more efficient product.

Lump sum: This is by far the most contentious and adversarial type of contractual arrangement—and the one most people associate with construction. A job goes out to bid, plans and specs are issued, a price is established with an agreed upon markup, and that's it—you own it. If something was missed, well, that's just too bad. Because after the contract is signed, it is not subject to renegotiation. It is up to each contractor to prove that what is being required falls outside of the original plans and specifications. That's why these types of jobs are flooded with change orders; this is the only means at the contractor's disposal to make up for estimating busts or job site inefficiencies. The mindset for the GC is that any deviation from the contract documents results in a change order with a value of four times the real cost.

Some owners actually prefer this type of arrangement, even though it means that the contractors are not necessarily working in the client's best interests. In the same way that the courts believe that they will eventually get at the truth by allowing opposing lawyers to tear each other's arguments to shreds, the owners believe they will get the true contract value by having the various entities cannibalize each other. The GMP route seems too loose to them. (In the public works arena, there is usually little choice. Most public work is done via lump sum.)

Unlike the GMP delivery system, the lump-sum environment locks the GC into accepting bids from the lowest responsible bidder. It's all about the price.

The temperament required in this environment—and there is really no polite way to put this—is to become an asshole as fast as possible. Phrases like "That's yours," "Tough shit, you bought it," and "Prove that's not in the plans and specs—I dare you" need to be incorporated into your daily lexicon. (This is the world of the grizzled veteran—not the soft newbie that just graduated from some tree-hugging university. Place the latter on these types of jobs at your own peril.)

A quick public works primer: The biggest mistake that GCs and subcontractors alike make on public works jobs is to assume that all of the various owners who are involved actually care about getting the job built. They don't. The sooner you realize that the only thing most public works people truly care about is how we are making them look politically, the better off you are going to be. This is not to say that there aren't some great folks in the public sector. But if they are efficient, effective, and willing to make decisions, they will soon become frustrated and quit, or will be forced out by their coworkers. In the public sector it is all about dotting your *i*'s and crossing your *t*'s and avoiding anything negative that could stick to them. Too often, GCs and subcontractors naively try to get ahead on the schedule so they can save the owner money, only to be told that they didn't fill out some seemingly trivial piece of paperwork correctly. When this happens, GCs and subcontractors often become angry and are overtaken with righteous indignation. All I can say is: get over it. Do what they said was required, per contract, and move on. (Even if you have to appoint one person to be the front man for expediting and bird-dogging only red tape, it is worth it in the long run.)

(As I mentioned earlier, no one gets fired in the public sector for saying no, but they can get fired for saying yes. One strategy that sometimes overcomes the predictable largess for the owner's own good, by the way, is to create cognitive dissidence, that is, making it uncomfortable for the owner to say no. For example, "I don't know. I guess we could drag this whole process out further. But I'd hate to be you when the powers that be find out that it was you who caused this delay." But if you use this strategy, do your homework first. Calculate the actual cost of the delay, and your willingness to attach account-ability and responsibility onto any particular government official, because they will never forget it. Therefore, don't make these kinds of decisions in a vacuum; make sure that your own upper manage-ment buys into this strategy. If you do decide to take this tack, it is vital that you pair their yes with a complete willingness to fully back them should the issue go south. And this is very important: be sure to heap on plenty of lavish praise should it go well. By mitigating the pain and increasing the gain, you'll actually help the public official in question feel more comfortable making decisions down the road. Is this manipulative? Damn straight it is. But it is for the greater good? Don't forget, it is *our* tax dollars—yours and mine—that are at stake here!)

Time and material: Time and material contracting is just as it sounds. Agreed upon pricing is determined for both materials and labor costs. Due diligence is required for the tracking of bills of mate-rial and labor. Your staff needs to be focused on tracking them both methodically, then determining the realism of the submitted invoices. Standard unit pricing and workflow rates are usually estab-lished for most construction activities and checked against what is invoiced. For example, the average electrician should be able to pull a certain amount of wire per hour. If what is billed doesn't come close to matching the actual work in place, then there is a problem. There is not a whole lot of art to this approach. The issues center on accuracy and shenanigans. For the contractor, if its crew goes faster than the average, while maintaining quality, then this is well-deserved money in its pockets.

Design-build: The advantage of design-build contracting is that it allows us and the subcontractors to assert a bit more control over the design and building process. Instead of the owner holding the architect's contract, we do. Problems arise when we don't work closely enough with the designer—or the designer is simply in over its head—because once construction gets ahead of the design, problems arise. (The director of the Denver Broncos stadium project, Emil Konrath, now of The Konrath Group, foresaw this problem and addressed it proactively. He made sure that he and his counterpart from the architectural firm had adjoining offices and that the bulk of the architectural staff were on site for all phases of the project, thus ensuring that the builders and the designers had constant access to one another. It was the primary reason why the project moved ahead so effectively even though market conditions at the time were less than favorable.)

There is a simple axiom in this modality: The A&Es are your friends! Besides, it is us, not the owner, who owns their work. You can't blame the architect in design-build work!

Integrated project delivery: IPD replaces the conventional design-bid-build model and is synonymous with Lean construction methods. IPD brings designers and builders together from the start. As described by Kate Moser:

> IPD calls for the project's team of designers, builders and owners to assemble early in the project's life and work collaboratively, sharing both risk and reward. Each member of the team can only succeed if the entire project is successful. But the method isn't altruistic; it also has legal implications as well. The legal basis of this new structure is a contract called an integrated form of agreement, or a tri-party collaborative agreement. (2009)

In May 2008, the American Institute of Architects (2008) released templates for IPD legal contracts. They define IPD as an "approach that integrates people, systems, business structures and practices into a process that collaboratively harnesses the talents and insights of all participants." Mutual respect and trust are fundamental, as well as the method's five key ideas: "collaborate, optimize, couple learning with action, and consider the project as a network of commitment and interrelatedness." Technology is also fundamental to IPD. Three-dimensional building information modeling (BIM) is frequently employed as a tool that allows earlier and more accurate cost estimates and facilitates greater collaboration among designers and builders. Applying lessons learned, reducing waste, continuously improving, monitoring results, and maximizing value, workflow management, and coordination of material handling are hallmarks of the approach.

The approach isn't for everyone, as it requires up-front costs and a great deal of trust (vulnerability). And many have only attempted to utilize it in naïve or superficial ways. But for those who have truly employed the method, they have experienced significant success. And I firmly believe that this is the model of the future.

BRIDGING STRATEGIES

Bridging strategies are tactics that are utilized when you need to get your team's attention and focus their energies in a particular way. They are helpful whenever you need to shift gears or raise a sense of urgency. They sound like the following:

- It's time to follow the letter of the law! Our contract is a killer and we've stubbed our toe because we haven't been clear about our scope. This afternoon, everyone is tasked with going back and reading their portion of the contract! From now on, there are no excuses for not knowing our scope of work and putting in place anything that hasn't been signed off.
- It's time to move forward! We've been hanging back accommodating changes, but we have to stop this way of thinking. In the long run, the owner will thank us for making this happen. Let's focus on building what is shown in the drawings. Look at your work, and the schedule and see what we can do to accelerate the schedule.
- Budget, budget, budget. Funding is tight as a drum. We can't afford any more mistakes. I'd rather we slow down and get it right. If you see something that doesn't make sense, by all means, raise your hands!

Feedback is pivotal when launching bridging strategies. Your staff will require frequent and immediate feedback as to how well they are making the transition from one strategy to another.

8

Lean Purpose

If we were at a party, and I asked you what you did for a living, how would you respond? If you're like most construction professionals, you'd probably say, "I build buildings," and quickly move on to the appetizer table. But let's say we'd both had more than our fair share of cocktails. If I pressed you, you might say something like, "I baby-sit subcontractors," or if you're a subcontractor, you might grunt and say, "I make those know-nothing GCs look good." Then it would be my turn to hit the appetizers. But let's think of the broader implications of this question. Most people in construction usually think of themselves as mere builders. And there's nothing wrong with that. In fact, there is something noble and humble about this belief. But whether you are a general contractor or a subcontractor, you are, in actuality, much more than a simple builder. You are in the dream business. Come on, put down the drink and hear me out for a second. This isn't just a bunch of philosophical hooey that I'm about to lay on you.

Think about the project you are working on now, and those you have worked on in the past; somewhere upstream, someone had a dream. Maybe it was for an office building that was environmentally friendly and featured new sustainable technologies. Or perhaps it was a high rise of condominiums set against the backdrop of snow-capped mountains with locally quarried granite incorporated throughout. Or maybe someone took a vacation, and brought back with him the notion of a deceptively nonintrusive shopping center that resembled a sunny hillside town from the heart of Tuscany. Or maybe, just maybe, the dream was much more down to earth, but no less important—a tilt-up warehouse designed to serve as a distribution center that would create hundreds of jobs, in a town where there had been few. Whatever the dream, the holder of it soon turned to an architect to capture it on paper. And what the architects delivered wasn't just a set of calculations. They came complete with schematics and artistic renderings that were designed to entice the senses and ensconce the heart.

So, as a construction professional, where do you fit amid all this dream weaving?

Believe it or not, front and center. You analyze the feasibility of the design, determine pricing based on the specifications, compare this with what the dreamer has to spend, develop a sequential building plan, and do your best to literally turn that dream into reality. Without the experience, analytical ability, and the check against reality that you are able to bring to the table, the dream could never metamorphase into its final form. It would remain just that—a dream. Now, I ask you, how cool is that? Most people spend their workdays doing the same boring thing over and over again. But you get to take something that was just an idea—a mere sketch on a piece of paper—and turn it into a tactile piece of functional art that literally transforms peoples lives.

The reason that this doesn't feel quite as magical as described is, given the role that you play in the process, you end up dealing with more than your fair share of disappointed dreamers. After all, who wants to be told that the lovely Cadillac-like structure that was drawn up for them by the architect can't actually be built without an additional outlay of cash or defying several laws of physics? When you have your heart set on a Cady, who wants to settle for a Chevy Cobalt—even if it is really tricked out? But that's the essence of what you do. You try to help people come as close to their dream as humanly possible as checked against market conditions and design limitations. And when it does finally become reality, because of your expertise, they can rest assured that the resulting building will stand the test of time. When I describe your job like this, doesn't it make what you do for a living seem pretty damn amazing? And doesn't it make all the discomfort, all the inherent conflict and the long hours that you log along the way, seem much more purposeful and livable? That's the point of having a sense of purpose. It helps sustain us during tough times.

Victor Frankl had much to say about the importance of a sense of purpose. At Auschwitz, he noticed that many people of fairly weak physical constitutions managed to survive the hardships of the camp far longer than did their more robust counterparts. He came to a key realization: There was an important psychological factor at play that determined whether or not someone survived. That key element was whether or not someone had something to live for. Perhaps it was their faith ("God commands me to survive!"), their family

("I want to see my wife and children again!"), or even a magnanimity of spirit ("I will not let these bastards take away my humanity!"), but whatever it was, everyone who survived had something significant to live for that helped them to endure the suffering that they faced each and every day. Conversely, fellow prisoners soon recognized that look in someone's eye that told them that he had lost his will to live. Often, they perished within days of making this psychological transition into hopelessness or what Frankel called "provisional existence," where the present lost all meaning (1959, 80).

These observations had a profound effect on Frankl. So much so that they led him to develop a completely new branch of psychotherapy known as logotherapy, which literally translates to "meaning therapy," that is, helping people understand the central role that a sense of meaning plays in their lives. As Nietzsche once said, "He who has a why to live for can bear with almost any how." (This may account for the wide difference in suicide rates between developed and underdeveloped countries. In developed countries the suicide rate is 13 per 100,000 people. In undeveloped countries, the rate is less than 6.5 per 100,000. The current thinking is that in undeveloped countries, people are preoccupied with the very notion of survival. Everything they do is toward that end and is therefore loaded with meaning. In developed countries, the notion of meaning as it pertains to survival is much more elusive and abstract. Things like lost status and an overall sense of alienation or pointlessness, which have nothing at all to do with survival, play a much larger role. Suicide is, evidently, the psychological "luxury tax" for living a much easier lifestyle.)

So what does all this have to do with being a leader of a project? I certainly don't mean to compare a job site with a concentration camp (though some of you who have worked for Heinz or Intel may beg to differ). But I believe that there is something that holds true for everyone when times are hard; teams that can endure the toughest owners, the most difficult contracts, and the most challenging site logistics usually have a leader at the helm who is able to evoke in them a strong sense of purpose. Maybe it is something as simple as establishing a sense of esprit de corp, a deep desire to never let a teammate down. Or maybe it is something a bit loftier, such as wanting to establish or grow a company or business unit from the ground up. Or maybe the leader is able to turn the difficulties the team faces on their ears and uses them as a challenge to prove that they can deliver under the toughest circumstances. Whatever the reason, as long as a team has a sense of purpose, they will endure. Without one, they tend to flounder, point fingers, and divide. They won't necessarily drift into oblivion and die like at a concentration camp, but several may quit or, worse, become an underperforming zombies (the psychological equivalent of quitting).

Let me give you an example of the power that a strong sense of purpose can have. I was hired by a contractor to resolve a dispute between their field payroll and IT departments. The people in IT complained that those in field payroll were overly hostile and tenacious when they didn't get their demands met. For their part, the field payroll people were incensed that the IT folks appeared unresponsive to their needs and seemed to exhibit a blasé attitude—often making them wait hours for what they viewed as fairly simple fixes. Clearly, neither side was caring a whole lot about giving their counterparts the benefit of the doubt. But there was something about this dispute that made it far from run of the mill. At its core, something extraordinary was going on. Early in the field payroll manager's tenure, she had pulled her troops aside when it appeared to her that they were just going through the motions. She said:

> Do you want to know something? When people look at us, they think that all we do is data entry, that we mindlessly plug a bunch of numbers into a computer program all day long. But let me let you in on a little secret: That's not what we do—not by a long shot. Do you want to know what we really do for a living? We help people pay their mortgages. When their parents get sick, we help them pay their hospital bills. When their kids are hungry or need new clothes, we help to feed and clothe them. Think about all the things that people depend on a paycheck for—that's what we do for a living!

She gave this speech three years prior, but it was burned in her staff's mind as if it were yesterday. So what was the upshot to this story? While I did ask the field payroll people to dial their intensity down a notch, you can bet that the last thing I wanted to do was dampen their passion. My message to the IT folks was simple: "The field payroll people will be more mindful of *how* they ask for things from you, and be more considerate of your time constraints, but they are not going to stop pushing for what they know is right. I suggest that you find a way to step up and tap into some of their passion." Unfortunately, the IT manager never took this message to heart and needed to be replaced a year later.

Now here is a key question to ask yourself: Do your people feel like they play a vital role in the overall success of the project? Does your receptionist understand that a lot of the opinions held about the entire job site staff and the company will be determined by how competently and politely he or she answers the phone and whether or not he or she redirects callers to the right people within the organization?

Does the document control person fully understand that everyone on the job—and I mean everyone—is dependent on him or her for easy and accurate access to the latest documents? Further, does he or she know that A&Es, owners, and inspector of records (IORs) will often judge the competency of the entire staff by how well the documents are maintained?

Per subcontractors, does every electrician pulling wire, every plumber laying pipe, and every sheet rocker placing drywall know the pivotal role they play in helping to get billings in place, which in turn is one of the key determinants as to whether or not a bank will extend a short-term line of credit to their company?

Believe it or not, all of the above statements are absolutely true.

Do you see what I'm getting at here in Lean terms? When people feel like they are a vital link in the chain, regardless of the job they do, they will embrace their role. As a result, productivity and accountability will go up, while waste and complacency go down. Conversely, when people feel like mere cogs in a wheel, they often do the bare minimum each day just to get by. Why? Because as human beings we are all hungry for meaning and purpose in our lives. If our work doesn't provide it, then our minds will drift toward something else. Maybe it's that football fantasy league that we just joined, or that new bicycle riding club that just started up, or maybe it's simply meeting up with friends for one more night of drinking. We fill the void of meaning in lots of different ways—not all of them productive. But when people feel like their work is vital, something interesting happens: they want to give their best every day because it means something to them and the people around them. And when someone wants to do something well, it also means that they are open to finding new ways to improve upon what they do. When we're bored, and have no hope to change our circumstances, we'll just do the same thing over and over and over again and won't look for ways to improve. I saw this mindset play out with workers at box plants all the time—until their supervisors found a way to get them involved. But when we have a sense of meaning, we actually search out new ways to be even more effective on our own, which is the very essence of a Lean culture.

So, if you haven't done it for a while, take the time to remind your people of the vital role they play on the project. Who knows, they just might make an extra dream or two come true.

But enough about your staff. What about you? What floats your prover-bial boat? What makes you look forward to getting up in the morning? I'll bet it isn't just a paycheck.

I remember sitting with Gus Sestrap (yes, the same Gus of "flow" fame) high up in the stands at the 50-yard line at the soon-to-be-completed Seattle football stadium project. With the finish line in sight, Gus was in a reflective mood. In the beginning, the project was beset with a great deal of conflict and infighting. But the management team persevered, and now the team was firing on all cylinders. Information was flowing freely between the field and engineering, and any conflicts that emerged were openly aired and resolved. Trust on the team couldn't have been higher, and victory was in everyone's sights. Finishing on time was not only *not* going to be an issue, but they were actually on target to deliver the job a week ahead of schedule—something rather unheard of for most stadium projects. Throw in the fact that the books were looking good and the owner was more than pleased, and what more could you possibly want at the end of a job?

So I asked Gus if there was a day when he knew that the team and the project had really turned a corner. He thought for a minute and relayed this story: They were about halfway through the project when he, the project manager (PM), the two senior project engineers, the two project superintendents, and the quality control manager were sitting around after lunch. By this time, a weekly lunch had become a ritual for the managers so they could talk about how things were going and recalibrate priorities. But out of the blue, Joe Lucarelli, the project superintendent, asked them all this question: "At this stage in your career, what is it that you like doing the most—I mean really enjoy doing?" They all thought it over for a moment.

"You know," Gus recalled replying, "I was just thinking about that the other day. I've been doing this for thirty years and have built just about every type of project that there is. I've realized that just getting the job done doesn't do it for me anymore. Don't get me wrong, I still love the challenge, but it's not what drives me. What I really get off on is watching the young people develop—sharing with them what I know and seeing them grow—that's what I really enjoy."

According to Gus, there was a lot of head nodding around the table. It turns out that each of them had been thinking about this same question on his or her own and had reached a similar conclusion. "At that point," Gus said, "a lot of the bickering for turf that had gone on really just melted away. Our discussions as managers went from 'I need this!' to 'How can we help our people get up to speed faster?'—a subtle shift, but a very important one." An unspoken competition suddenly arose between them. They began trying to outdo each other for who could develop the most people the fastest. The unintended payoff on all of this was huge. The team truly began working together—from top to bottom. Information, coaching, and mentoring flowed freely. By the end of the job, the unsure kids with tiny voices (and there were about thirty of them), over time, became lions with confident roars.

And there was one more unanticipated benefit. Unlike most jobs, where at the end, the managers lay spent like the Spartans at Thermopylae, from all the work they had to take over from their "incompetent" juniors, this management team was surprisingly spry. By helping the staff become more competent, the managers had actually made their own lives easier. Rather than standing alone as "heroes," they were about to cross the finish line with an incredibly well-developed group of underlings who were not only *not* in need of rescue, but who were poised to take on the responsibilities of their next assignments. At the end of the job, the only thing people felt truly bad about was knowing that it was unlikely that they'd ever have the opportunity to work together again.

So, who says philosophy is just a bunch of hooey?

Opportunities for establishing a sense of purpose abound and often start with the small stuff—right down to the paperwork. I don't know about you, but if I understand the importance of why a certain piece of paper needs to be filled out, I can do it all day long. But if filling out that same piece of paper seems meaningless, I won't do it even if you stand over me with a can of gasoline and a match.

So, the next time you are having difficulty getting people to adhere to a specific policy and procedure, or your team is simply in the midst of a midproject malaise, take a step back and try a little experiment. Take the time to explain *why* what they are doing is so important. I'll bet that you'll see a spike in their performance as a result.

9

The Conflict Paradox: Encouraging Debate without Letting It Become Destructive

At the risk of dating myself, I wonder how many of you remember the *Twilight Zone* episode entitled "It's a Good Life," in which Billy Mumy plays a little boy named Anthony who possesses special powers to do just about anything he wants. Being an omniscient child, the things he took delight in were, in equal measure, facile and horrifying. Mirroring his insecurities, he demanded complete and unwavering assurance from the adults trapped in his unwitting web of tyranny that everything he did was the most marvelous, beautiful, funny, or intelligent thing that any of them had ever seen. And if anyone questioned his actions or broke under the unrelenting pressure of maintaining this delusion, they were either transformed into some pitiable deformed creature or "sent to the cornfield." We never do find out what the threat of being sent to the cornfield actually means, but by his subjects' terrified and sycophantic exclamations of "That's a good thing that you done, Anthony, real good!" each time he does something terrible, we know that it must be something truly awful.

The story was a metaphor for a society suffering under the yoke of a capricious, foolish, and tyrannical leader. The episode was based on a short story written in 1953, when memories of Hitler, Mussolini, and Stalin were still fresh in everyone's minds, and thoughts of why so many seemingly good people were capable of capitulating in so many grievous acts gave everyone pause. Years later, psychologist Stanley Millgram was able to replicate these findings in the laboratory and provide us with some answers. He discovered that, in the presence of a stern and commanding person of authority (in his case, a forceful university professor), his subjects continued to deliver a series of progressively stronger electric shocks to what they believed were fellow participants, even when they reported severe pain or heart-attack-like symptoms. According to Millgram, when we are under the influence of a powerful leader, most of us are likely to conform our behavior to his will—not because we believe he is right, but because we feel powerless to defy his authority.

Though not as prevalent as in days past, I've seen this same pattern of behavior play out on job sites. Convinced that the best way to run a project is to seize control of every aspect of the job, a strong-willed leader, through the force of his will, begins to drive the project—and in the process, squelches everyone in his path. By dismissing questions and mocking ideas and input, or through sarcasm and withholding bits of key information, he ensures complete obedience to his plan. As he is usually the most experienced person on the job, and seemingly controls people's fate by being in charge of writing their yearly reviews, those under him meekly acquiesce to his authority. Even those with a modicum of experience find themselves either tacitly agreeing or holding their collective tongues for fear of being banished to the construction cornfield. ("How's that tilt up in the Badlands of South Dakota sound to you?")

Sadly, the very thing that should lift the team to greater heights—the leader's experience—is instead wielded against them like a mighty sword. As a result, the team grossly underperforms—doing little more than what is doled out to them in drips and drabs, and hating every minute of it.

In most cases, creating such a dynamic isn't the leader's true intent. Though there are those for whom being in complete control is a psychological obsession, most of the time, the person exhibiting these behaviors is simply an overly responsible individual who is so overwhelmed by all that needs doing, and so befuddled by how to get there as a team, that he resorts to despotism out of a feeling of sheer desperation.

Compounding this problem is the fact that most people in construction are heavily lauded early on in their careers for the very behaviors that make them such terrible leaders down the road—their ability to exhibit control and dominance of those they are tasked to oversee. And what we know from years of psychological research is this: when under stress, people will revert to behaviors that they believe brought them their initial success—and they will do so with a vengeance. If they believed that, in the past, a good result was attained through a little yelling, they will now yell their bloody heads off; if they thought being a little controlling was of true bene-fit, they will quickly transform themselves into the ultimate control freak.

Unfortunately, managers who rely on control, bullying, and manipula-tion as their primary management tools rarely experience a smooth path to their intended target. Why? For two primary reasons: Because they are so results driven, they are oblivious to the collateral damage in the form of resentment, resistance, and hostility that emerges in their wake. And second, because the seemingly willing often feel backed into promises that they have no intention of keeping, leaders inadvertently set up a situation, as described by Macomber, where any promises made by teammates are com-pletely unreliable. When under duress, what should be a firm no becomes an "uh, sure." Why are these unreliable promises such Lean killers? It is because the leader mistakes a weak yes as buy-in, even though the task will likely be dropped. And when the task is dropped, this, in turn, increases the leader's belief that the only way to get things done properly is to either do it himself or be even more controlling, which creates more fake acquiescence, and yet another self-reinforcing recursive cycle is born.

I can picture a number of you experiencing a little tightness in the col-lars right about now. If what I've described sounds painfully familiar, and you are not happy about it, ask yourself this question: Is there a way of getting where you need to go without resorting to controlling the heck out of everything? Believe it or not, there is.

If you honestly believe that you are the only person on the job who can get things done properly, then you have another problem on your hands, don't you? Either you are ridiculously overstaffed—and are taking a need-less hit on your staff-to-return ratio—or you're systematically distorting the way you are looking at yourself and the people around you. I'm think-ing it's probably the latter.

So let's try something else. Think back to our discussion about individual ethics and the idea of what it takes to be a good PM or a good field person. What is the best way to help instill this way of thinking in your people? Believe it or not, in construction, people learn best not by receiving a series of edicts from their leader, but by being able to test and play around with ideas. I've talked to thousands of you over the years and have heard many tales of how, when you were kids, you compulsively took apart anything you could get your hands on and put it back together again. These behav-iors laid the foundation for who you are today. By deconstructing then reconstructing, you began to understand what it took to build something.

It is by a very similar mechanism that people at job sites learn to develop their skills. Through the act of discourse (them asking questions about procedures and methods, you responding to their queries) they have the opportunity to deconstruct and reconstruct concepts in their minds. By asking questions they are not questioning your authority or abilities; your staff is actually "playing" with construction ideas, that is, measuring them for accuracy, validity, adaptability, and fit. More importantly, it is through such dynamic play that they gain a true understanding of how to think about their work. And all of this is at the heart of developing a Lean culture. Believe it or not, we actually *want* people to constantly ask themselves and others, "Why do we do it this way?" Such questions lead to a much deeper understanding of the principles behind the task and, in turn, clear the sightlines for seeking potential process improvements.

And here is another reality about how people acquire knowledge: most people learn best when they are allowed to make a few mistakes. They don't learn as well when they are dictated to and controlled by the "smartest" guy in the room. I know this runs counter to what most construction leaders believe is acceptable. And yes, I am well aware that mistakes cost money. But mistakes are also valuable learning lessons. If we want our people to develop, we have to allow them to make some of their own decisions and run with them. Unlike tyrannical Anthony, we have to become less enamored with our own shining star and promote an atmosphere where it's okay to voice an opinion that runs contrary to our own and allow for some mistakes along the way versus always striving to make sure that nothing bad ever happens.

Here's a quick little assessment to see how well you are doing in terms of creating an environment of discourse. If people line up outside of your office door to ask a quick question or two, and then quickly scurry off, that is not discourse. You are simply holding court. Discourse means that an actual interaction occurs—characterized by a give and take of ideas.

Another measure is whether, privately or in meetings, people speak up and challenge you on your approaches. I'm not talking about the belligerent, self-serving little piss ants who are just trying to make a name for themselves by cockily challenging your every move. We'll talk about how to address these folks later. What I'm referring to are those staff members who are honestly trying to wrap their arms around where you are going. It's these folks that you most want to encourage. Effective leaders make room for people to ask probative questions in order to gain a deeper understanding of their direction.

Therefore, if no one bothers you all day, if your staff meetings are as quiet as Holy Communion, or if you only hear about bad news well after it has hit a crisis point, you'll need to change some things around. The last thing any leader wants is for his or her staff to resort to veiled acquiescence or highly filtered feedback for fear that what they say will be used against them. Remember, if we want to develop a true Lean culture, people need to be able to speak openly and in an unguarded fashion.

Here is how you can encourage what Lencioni calls "healthy and productive conflict"—and without a whole lot of effort. Either in meetings or one on one, simply ask people what they honestly think about a particular issue (particularly the quiet ones) and thank them publicly for putting their ideas forward. By doing so, you will get more of the same and less of the silence that is so Lean killing. And by working hard to squelch your impatience ("That's the dumbest thing I've ever heard!") and substituting respectful consideration instead ("I considered that, but let me explain why I think it's better if we go another way"), you'll send a strong signal to the team that you value their opinions, even if they run contrary to your own. As a side benefit, you may just happen to hear a really good idea that you hadn't thought of on your own.

This means, of course, that you can't do all of this through clenched teeth. Remember, people will pick up on whether you mean it or not. And for your own sake, I want you to mean it. It is through this type of productive conflict that your staff will learn and grow—and you will avoid an early heart attack. Because when they grow, you'll be able take some of the things that are on your plate and actually move them onto theirs.

For those of you of a less tyrannical bent, but who are nonetheless puzzled as to why your staff meetings are so silent, there is something else to consider, and this is particularly true for those of you working west of the Mississippi or with a very young crew. These folks have come to believe that any level of passionate discourse is somehow bad. Blame it on the 1960s, or a more feminized, nonconfrontational approach to communication, but when people within these demographics hear the slightest raising of voices, they run, at least psychologically, for the proverbial hills. If you think it's difficult here, try the Netherlands. There, people on job sites can actually call the police if their boss raises their voice at them. I jokingly said to a VP who was from the East Coast that if this same law applied in the United States, New York and Boston would quickly become penal colonies. He wasn't all that amused.

This is unfortunate, because the baby is being thrown out with the proverbial bathwater. Conflict, when healthy, is a tremendous source of creative ideas.

Paradoxically, the same people who run from passionate debate aren't shy about expressing their dissatisfaction through more passive-aggressive means, such as intentionally doing things that they know will make others angry, like not returning phone calls or emails, or registering formal complaints with HR. All the more reason why we want to encourage the expression of dissenting views openly and honestly. The more we make this a standard operating procedure, the less likely it is that they will be expressed via more subversive means. Therefore, it is incumbent on us to help dissatisfied workers channel their passive angst in more productive ways and see the value of engaging in productive conflict. That means that we have to work harder at soliciting people's input, encouraging them to keep going when they do take the risk to express an opposing view, and highlight the positive outcomes that are a direct result of passionate discourse.

On the flip side, this also means that we have to be willing to confront people when they voice their displeasure through gossip or other indirect means. While a job site isn't like the old Soviet Union—we can't dictate what people can and can't talk about—we can point out the benefits of creating an environment where people can speak openly without fear of retribution, including that of being talked about behind their backs, by encouraging open, honest, and direct airing of frustrations.

So in summary, by making room for and encouraging people's questions and ideas, by praising them when they voice dissent in a professional manner, and by taking the time to work through differing opinions, we open ourselves up to a whole new world of ideas, solutions, and team interactions, and that is a good thing—a real good thing.

Come on, be honest. Deep down, don't you want people to commit to a course of action because they actually think it's the best idea out there, rather than simply because you said so?

RESOLVING UNPRODUCTIVE CONFLICTS

What do you do when the run-of-the-mill disagreement is no longer run of the mill? People are hot, they can barely stand to be in the same room together, and each is thoroughly entrenched in his or her own position.

Does this mean you have been a bad leader and all is lost? Hardly. Perfectly good teams, with perfectly good leaders, can, from time to time, become temporarily derailed. But how you handle such derailments determines whether or not your team gets back on track or remains in a ditch.

The greatest strength of construction people is that they are not dead from the neck up. Most hold passionate views about what is the right, what isn't, and what should be done about it. But precisely because of this quality, their greatest strength can quickly turn into their greatest weakness. Often, they become entrenched in their own positions and have difficulty making room for viewpoints that aren't their own. Obviously, nothing impacts the flow of a job more than teammates who go out of their way to actively avoid another, or who wittingly or unwittingly sabotage one another's efforts. It is, therefore, important for you to keep your balance amid this turmoil. Good leadership isn't about avoiding these types of conflicts; it is about harnessing their passion so it will remain productive (versus turning destructive).

Before discussing how to handle these types of situations, let's focus first on what not to do. Never make consensus your goal. Consensus is the destroyer of great ideas for the sake of what Patrick Lencioni calls "artificial harmony." It takes a choice piece of USDA prime porterhouse and grinds it into mushy meatloaf. Consensus is the acceptance of mediocrity or wrongheadedness in exchange for interpersonal comfort. A great case in point is the Groningen Art Museum in Holland. Let me explain just how this monstrosity came to be. The city council of Groningen, being a collectivist sort, decided to include input from the community for the design of its new art museum. Mind you, this wasn't to be some paltry lip service. The community would have a real say in determining the eventual design. As is usually the case with such pie-in-the-sky thinking, it didn't take long for all of this to become marred by the ugly fact that there is rarely enough unity in any given community to come to an effective decision on anything. It wasn't long before hard-line factions formed. One group wanted a traditional, easy-to-maintain design; another wanted a distinctive, ultramodern radical design; while another favored something in the middle. Passionate debate was had, merits of each design were weighed, and nothing at all was settled. So in their infinite wisdom, the city council decided to go with a little bit from all three designs. The picture in Figure 9.1 is a testament to what happens when consensus seekers are given free reign to run amuck.

As it turns out, the visual design was the least of this project's problems. One faction had become so enamored with the idea of building the

FIGURE 9.1
Groningen Art Museum.

museum on a piece of land that jutted into the river that they managed to find an engineer who, despite many naysayers, told them that it indeed could be done—and for an acceptable price. Concerns for potential flooding were dismissed as a once in every two-hundred-year possibility, and the plan was adopted. Unfortunately, no one foresaw that this possibility could happen a year after the project's completion, but it did. The hapless curator spent, along with the rest of the museum staff, much of that winter wading through calf-deep water hand-carrying valuable paintings and sculptures to higher ground.

So, what is the moral of this story? What you want for your project is the *best* ideas, not little pieces of good ones that together don't amount to much. Compromises rarely will take you where you need to go. At some point you will have to make a decision between two competing options that make the most rational sense. For example, when faced with the choice, you can't run electrical conduit both below ground and in raceways. It has to be one way or the other. But first, you will need to frame them as distinct options, or as Roger Martin eloquently puts it in his book *The Responsibility Virus* (117), "A well-framed choice is an irreversible commitment." So how do you effectively make these kinds of choices?

Let's take a hypothetical situation and run it through a suggested framework. Let's say that you are a project executive (PX) in charge of a large project, and your management team (the project manager, project super-intendent, lead project engineers, and project accountant) is having dif-ficulty coming to an agreement on how to adequately track and report costs. In fact, some discussions in the past have become so heated that one or more parties walked out of the room and did not return. Worse, you've heard rumors that the engineering side of the house believes that they are being set up to be the scapegoats when the project eventually fails. And you're only two months into a three-year project!

So how do you handle this situation? First, it's vital that you set a new tone for the upcoming meeting that you are about to have. When all are available, gather the pertinent parties together in your office for an impromptu announcement:

> Thanks for coming. You're probably wondering why I have gathered you together. Tomorrow at 2:00 p.m. we are going to meet as a management team to discuss our budget tracking and reporting issues. This meeting is manda-tory, so clear your calendars. Here's the deal: we need to come up with an agreement, and none of us leaves the room tomorrow until we do. This issue is too important for us to leave it hanging any longer. People below us are depending on us. We need to set an example, both procedurally and in terms of teamwork, that the rest of our team can follow. That's it. I'll see you tomor-row. Please come prepared with good ideas—and an open mind.

In the meeting, do the following:

1. Establish common ground that everyone can agree on.

   "I think I'm stating the obvious when I say that we are at an impasse in terms of how to track and manage the budget. And I think we'd all agree that we'd all feel a lot better if we could form an agreement and get past this."

   (It's important for people to know that there is at least one thing that they can initially agree on.)

2. Get people to state what is at stake for them *personally* if this conflict continues without resolution.

   "I'm going to be completely honest with you; I get a rock in my gut each time I see one of you approach my office knowing that you are coming to complain about this issue. All this bad mouthing and ill feeling is, quite frankly, making me miserable. I'm starting to take it

out on my wife and kids and I don't like it. I'm just curious, is anybody else experiencing what I'm experiencing?"

(Make everyone answer this question. Once people recognize that they are not alone in their suffering, they are much more willing to listen to opposing points of view.)

3. Brainstorm ideas.

   "Look, right now, all I want to do is get ideas up on this white board in terms of how to solve this thing. I don't want anybody to judge whether the idea is good or bad at this point—we'll sort that out later—I just want to get them up here."

   (Write down each idea. Resist the temptation to edit or begin to favor one idea over another. If you do so, you will truncate this process before it starts. Make sure you prod the quiet ones. Everyone has to participate. You don't want anybody leaving the room feeling like they never got the chance to speak—thus giving themselves ample justification for not following what is subsequently agreed upon.)

4. Look for common themes.

   "Are any of the ideas that we've put up here similar?"

   (If you can, have the team group the ideas into categories. Often, people don't realize that they aren't as far apart as they think they are.)

5. Start the idea selection process.

   "Without taking into account who said what, are there any ideas, or a combination of ideas, that are starting to grab you? If not, do we need to think of something else?"

   (Usually, at this point, traction starts to take hold; an idea or two starts to emerge from the pack.)

6. Design a test.

   "Okay, so it sounds like we've settled on how we are going to track costs. And at first blush, it looks to be in line with what the owner is looking for from us. But let me play devil's advocate for a minute. How do we know we've come up with the best idea? How would we test it?"

(This is an important step. Coming up with a way to test the idea builds people's confidence in the agreement. It reassures those who were on the fence, and keeps people focused on analyzing the issue objectively.)

7. Analyze.

"So, it sounds like we are all agreeing that the best beta test for what we've come up with is to run it through our Prolog system. The next step after that is to see if it flies with our accounting system and with what the owner is expecting to see. I think one of the reasons we got into this mess is because the owner has some pretty unique real-time reporting requirements that our systems really aren't set up to deal with."

(Allow others to chime in and refine. At this point, people will start to demonstrate a willingness to commit.)

8. Summarize.

"Okay, sounds like we are there. Now this is important. Can anyone summarize what we came up with?"

(Before you move forward, make sure that *everyone* is on the same page.)

9. Commit.

"I want everyone to look each other in the eye and make a commitment that this is what we are going to do. No one gets to go "Maverick" on us. We do this together or not at all."

(You have my permission to be very harsh should anyone unilaterally violate step 9! No one gets to break this agreement unilaterally, including you!)

10. Commit to follow-up.

"Look, down the road, we may find that this plan isn't working because some unforeseen factor came into play. For instance, I'm hearing rumblings already that the owner may want to tweak the reporting requirements. Let's meet in three weeks to see how this plan is working and to see if we need to modify it in any way. But this is important, and I can't say this strongly enough; if it turns out that we have problems, no one makes changes unilaterally—we do it together—agreed?

And it is important to add…

"Thanks, everyone. It means a lot to me that we were able to get this done as a team!"

If you adhere to the above steps, you should be successful. I've yet to have this format fail. But there are a few ways that it can go awry, so try to avoid these common pitfalls:

- The leader allows the process to become personal. (Nobody is allowed to call anybody else stupid, thickheaded, or a dirty motherf_____— even if they believe it to be true.)
- The leader advocates for a particular position too soon in the process. (Everyone has to feel heard or this just doesn't work.)
- The leader allows the most forceful or aggressive people to dominate the discussion. (Aggressiveness does not necessarily equal the best idea.)
- The leader allows people to walk out of the room and not come back. (People can take time to cool off, but they cannot just leave and not return. I actually consider this a form of job abandonment.)
- The leader doesn't call people out should they breach the agreement. (If you fail to do so, you are giving tacit reinforcement to those who break their word.)
- The leader doesn't follow up to see if everyone still thinks the agreement is working. (The ability for a group to fine-tune and recalibrate is central to any successful plan—and is part and parcel to Lean thinking!)

That's it. Don't be afraid of conflict. Conflict can be your best friend. If addressed up front versus being allowed to fester, it can be the wellspring of great ideas. And, believe it or not, you don't need to be a psychotherapist to work through them! It is just a matter of sincerely gathering your ethics, listening to all points of view, and keeping the team focused on finding the *best* solutions. Do these things well, and by the end of the project, the team will actually forget that any serious impediments had occurred at the outset!

10

Establishing and Maintaining High Standards

Here is an oft-asked question: "If we're going to focus so much on teamwork, aren't we going to sacrifice quality?" Here's my answer: if you misinterpret working as a team as merely getting along, you will. But true teamwork isn't about simply holding hands and singing campfire songs; it's about working efficiently and effectively together in order to produce a desired result, which, by the way, may or may not happen to include getting along.

Equating teamwork with getting along is an easy enough error to make. After all, we've all had plenty of experience living with atrocious behavior for the sake of keeping the peace. In California, even in fine restaurants, it's not unusual to encounter a bevy of young children yelling and running about as their parents contentedly eat their meals, seemingly oblivious to the annoyance that their children are generating. So what do the rest of us diners do? Do we ask the parents to get off their posteriors and get their children under control? No, we don't. We ignore the commotion as best we can, pay our check at the first opportunity, and leave. Why do we put up with it? Because in California, getting along is prized over everything else. In fact, to speak up would be considered an ugly display of intolerance—even though the only thing we'd actually be intolerant of is bad manners and poor parenting skills.

Besides revealing what a middle-aged curmudgeon I've become, what does this little rant have to do with leading construction teams? To some degree, we've all learned to associate lax standards with a veiled, albeit artificial, sense of togetherness. But for our teams to truly be Lean, standards need to be set—not just in terms of how much work is to be produced, but also for how people are expected to work together.

Frankly, I've seen teams who have gotten on wonderfully but weren't worth a damn in terms of the product they produced. When they should have been engaging in productive conflict or pressing their teammates for promised deliverables, they held their tongues instead. They did so because their priorities were centered on preserving their friendships versus producing the best possible product. This is *not* a Lean culture, and should never be the priority for any team that genuinely values its customers.

Conversely, I saw one project where people didn't get on particularly well (50% were vocally prochoice, while the other half was adamantly pro-life), but they consistently set aside their personal differences to rally around a unifying set of goals.

Let me share a few of stories that I hope will make the distinction between simply getting along and that of truly working as a functional team even clearer, as well as demonstrating what leaders can do to bring about these crucial differences.

I received a call from Rob Stein, VP and operations manager for KCS West. He had just staffed a new project and there was friction from the start. The project manager who was in place had a great deal of construction experience, but was new to the company. For his part, the project superintendent had ample technical ability and abundant knowledge of company policies and procedures, but interpersonally was a bit of a prima donna. The superintendent had a history of working on limited-scope, stand-alone projects, where he received minimal supervision. But this project was a horse of a different color. Scopes were broad, the dollar amount was substantial, there where multiple staff in place, and the job would require a high level of coordination, as the architect on the job was in well over her head. Rob knew that if this project was going to be a success, this team was going to have to pay particular attention to the drawings and properly document every design issue *before* any activity was implemented in the field. In short, the engineering and field functions would need to be in lockstep with one another. Unfortunately, from the moment the PM and PS met, the superintendent seemed to go out of his way to make sure that he and the PM would not be walking the same path. Rather than helping him learn the ropes, the PS took every opportunity to point out the things the PM didn't know about company policies and procedures, often doing so in front of the other staff. To say that the PM felt undermined is a bit of an understatement. To top things off, the PS also made veiled threats about quitting the company if forced to remain under such an "incompetent" PM. As you can imagine, this was more than a little awkward for everyone involved.

Most people in Rob's position would have made the mistake of allowing this type of situation to play out before intervening. But not Rob. He called the PM and PS into his office and promptly informed them that he would be making me available to help them work out their differences, but that there was to be no misunderstanding, they weren't being asked to *try* to work them out. They would either find a way to work together or changes would be made. And he made the point of emphasizing to the superintendent, "If I do need to make changes, let me be clear, I'm not removing the PM." By taking charge of the situation early on and making it abundantly clear what he expected, Rob sent a clear message about teamwork and high standards. What he conveyed was

- Working as a team was not optional.
- No one person is above the team.
- There will be no payoff for undermining anyone.
- No one's ego is more important than accomplishing team objectives.
- No one individual will be allowed to hold the project hostage or be a barrier for achieving team results.
- People will have an opportunity to work things out for themselves, but if they choose not to, things will be worked out for them—perhaps not to their liking.

Rob is the kind of leader that I love working for because he knows exactly what he expects from his teams and he is willing to back it up by doing what most people tend to avoid like the plague—calling people out on their team killing behaviors

Now here is the happy ending to this story. For all of his posturing, the superintendent was, deep down, a pretty ethical guy who was experiencing a significant spike of insecurity about having to work directly under someone for the first time in six years. But to his credit, as the PM clarified what he expected and needed from the PS and, more importantly, what he *didn't* feel that he needed to be involved in, the PS listened and dropped his guard. As each took turns putting their worries and concerns on the table, they each discovered that they had far more in common in terms of how they envisioned executing the job than they initially believed. At this point it was clear that they were no longer going to have a problem working together. And for the life of the project, their relationship held up well. As for the underperforming architect, that's another story entirely.

Here's another example. I had known Ken Schroeder when he was a project executive (PX) with another company. He had subsequently moved on and begun working for Blach Construction in a similar capacity. Three years went by before we were able to catch up. When we met for lunch he was clearly very happy about his new situation. When I asked him about his company's core values and if this had anything to do with why he enjoyed working for Blach so much, he thought for a moment and then smiled. "That's an interesting question," he said. "For one, morale here is very high. We have very high standards about our work product, our work ethic, and most of all, integrity. People are given a reasonable period of time to learn the culture and what is expected, but if, after a year or so, it isn't a good match, they are invited to leave the company." When I asked him why he thought high standards led to such high morale, he said something quite revealing. "At Blach, when you pass the test, and are fully part of the team, there is no second guessing. If you have an idea, people don't say, 'well, you'll have to run that by so and so first.' They simply say, 'Ken, we trust you—go for it.'"

This is the often overlooked benefit of establishing and adhering to high standards—they build trust. People don't have to be managed around, manipulated, cajoled, or tolerated. They are trusted because they have already demonstrated that they can hold themselves accountable to the standards and manage themselves accordingly—no micromanagement is required in such an environment!

Here is another situation from a subcontractor's perspective. Steve Foxworthy, a VP of field operations for Rosendin Electric, fiercely confronted a PM after his job posted a substantial financial loss. Did he take this young man to task because his job lost money? On the surface this would seem to have been the case. But in Steve's mind, the issue ran much deeper than that. What stoked his ire was the fact that the PM chose (the use of the word *chose* here is intentional) not to follow the company's established policies and written procedures. The PM knew the job was in trouble and *chose* to bury it in the false hope that he could somehow magically pull the job out of the fire without anyone noticing. But this was exactly the point that Steve knew he'd have to press home if this PM was ever going to have a chance of being successful at Rosendin: if the PM had simply followed the company's procedures, the problems would have been flagged early on, and the company's upper management could have mobilized its resources to help right the situation immediately. But by choosing a selfish course of action (hiding the weenie in a vain attempt to keep himself from looking bad), the PM prevented the company from helping him and, as a result, needlessly lost a substantial amount of money. Again, it wasn't about the money, per se. It was about the PM losing sight of what the standards represented and why these protocols and safeguards were established in the first place—to ensure overall project and company success.

The best thing about this story is that it didn't come from Steve, but from the PM in question. And he didn't relay it as some "woe is me" tale. To be honest, I don't think I've seen anyone so disappointed in himself in my entire life. The chagrined PM said, "Man, I felt like I had just let down my dad." More importantly, he had a true *Hansei* moment. He fully understood the impact of his actions and had a plan for making amends. "I intend on working with this company for the next thirty years. Believe me—I will never do that again! As much as it hurt to hear what Steve had to say, it made me proud to work for a company that has people like him who actually take the time to tell me when I've screwed up. Most places would have just fired me. As upset as he was, I could tell that Steve was pulling for me to succeed. And it's good to know that as long as I trust the company and don't pull another bonehead move like I just did, I will."

Unlike Steve Foxworthy, when left to our own druthers, most of us would rather ignore bad behavior than confront it. But here is the reality: what we choose to overlook we choose to live with. We can blame our bosses for sticking us with a lousy contract or rant about not receiving the proper support, but when standards slip, the truth is, it is largely by our own doing. Here are a few "look in the mirror" realities to keep in mind:

- When we choose to ignore poor performance, in reality, we are choosing to accept mediocrity.
- When we choose to ignore displays of contempt by one teammate toward another, we are actually choosing to allow trust on our team to be destroyed.
- If we choose to not take the time to share what we know, or check in to see whether or not what was shared was actually understood, in reality, we are choosing to allow others to guess at solutions rather than know them.

- When we choose to dole out tasks rather than put them in context, we are choosing to have automatons and functionaries on our project, rather than thinkers.
- When we choose to drop down and do the work of others, rather than holding them accountable, we are choosing to give up our role as leader in favor of being a doer.

What we get from our teams is more up to us than we realize, which is actually good news. It means we have more control over getting what we want than we think. The more we help our teams to improve, and demand nothing less than their best in return, the closer we are to getting the product that we actually need.

This also means that we have to be willing to fight for what we want. I've seen teams go down in flames simply because the leaders weren't willing to put their money where their mouths were. Here are the things that every leader should be willing to fight for or defend:

- That everyone on the team should be able to speak their minds without fear of ridicule or worrying that what they said will be used against them at some future date.
- That when they are upset, team members will speak directly with those who they are upset with versus talking to everyone *except* the person they have the problem with.
- That if there is any dirty laundry on the team, it will be washed in-house.
- That work issues will remain work issues, rather than becoming personal. Objective solutions will be sought versus obsessing on how to assign blame.
- That everyone is expected to share what they know.
- That it's everyone's job on the team to make all of their teammates successful—no exceptions.
- That no individual on the team gets to unilaterally decide which goals will succeed and which ones will be allowed to fail. If a goal is perceived to be in jeopardy, it is to be identified, discussed, and jointly rectified.
- That any resulting course corrections to achieve a goal will be communicated to the rest of the team.
- That everyone will readily share resources to make sure that no team goal fails.

- That everyone is expected to give their best in accordance to their capabilities—no exceptions and no excuses.
- That if something comes up in someone's personal life that intrudes on his or her ability to give his or her best, he or she will let his or her teammates know and ask for help.

If any of these principles are violated, you should be prepared to take people (calmly and professionally) to the woodshed. After all, what's the point of saying that teamwork is important if you aren't willing to fight for or defend it? We need not lose our humanity in the process, but we do need to confront issues head on and assert what we expect to be done differently.

There are two questions that we often fail to ask when confronting our teammates on their bad behavior, and ironically, we usually fail to ask them of our highest performers. They are

"What's freaking you out right now?"

"What help do you need from me so you won't have to resort to _____?"

We sometimes forget that our best performers are also human and that they too have a finite capacity. Sometimes in their desire not to disappoint, they take on too much or resist saying "no mas" (no more)—until they break or take their stress out on somebody else. Asking the above questions amply conveys that you are giving them the benefit of the doubt about their bad behavior without making any excuses for it. By putting their behavior in context, you are taking the time to listen for the reasons behind it, while still demanding that the unwanted behaviors stop.

Whenever you see changes in behavior for the worse, be sure to take your top performers out to lunch immediately and find out what's going on. But don't let them off the hook too easily. After all, they are not just responsible for producing estimates, cost reports, or bills of materials. They are also accountable for promoting teamwork. The goal for everyone is to do their part to keep the line moving efficiently and effectively. But given their track record, go the extra mile to gain their perspective. It will be the best way to help them to regain their balance.

One last point about high standards, and it is counterintuitive to the way that people in construction have been brought up to think about them. Most people in this business are taught to go, go, go!—that time is money, and a good day's work is predicated upon going as fast as possible at all times. While it is true that time is money, so too is the waste generated by inadequate planning. Toyota employs the concept of *Andon*, and it is central to their commitment to continuous quality improvement (*Kaizen*). Any time an employee on the line notices a quality issue that he cannot solve himself, he is *expected* to pull a handle, shut down the line (*Andon*), and ask for help. In fact, managers at Toyota are often confronted if the assembly line runs too long *without* a worker shutting it down. The assumption is that the manager has allowed employees to become complacent, that is, favoring production over quality.

Please encourage the people you work with to call a time-out, throw the flag—or any other symbol that you want to invoke—whenever they feel confused, unclear, or have been given competing directions by various managers. As a true Lean leader, these are the times when you should shut down the production line and gather the troops so that a chain of replicated mistakes won't be formed. I don't know of any manufacturing company worth its salt that values speed of production over quality, yet in construction, this standard is violated with alarming frequency.

Let's step away from construction for a moment and examine how the failure to employ the concept of *Andon* can have truly devastating consequences. On March 27, 1987, in a heavy fog, KLM Flight 4805 roared down the runway at Tenerife Airport, and struck Pan Am Flight 1736 as it was taxiing in the opposite direction down the same runway. It would be tempting to blame the accident on weather conditions and a confluence of other factors. But the real culprit was a failure to adhere to high standards and the unwillingness, at all levels, to "shut down the line." Below is an analysis of the communication errors that transpired prior to and during takeoff (from *Wikipedia* and Krause, 2003, 199).

Immediately after lining up, the KLM captain (van Zanten) advanced the throttles slightly (a standard procedure known as spin-up, to verify that the engines are operating properly for takeoff) and the copilot advised the captain that air traffic control (ATC) clearance had not yet been given. The captain responded, "I know that. Go ahead, ask." The copilot then radioed the tower that they were "ready for takeoff" and "waiting for our ATC clearance." The KLM crew then received a clearance that specified the route that the aircraft was to follow after takeoff.

The instructions used the word *takeoff*, but did not include an explicit statement of whether they were cleared for takeoff.

The KLM copilot read the clearance back to the controller, completing the readback with the statement "We're now at takeoff" or "We're now, uh, taking off" (the exact wording of his statement was not clear), indicating to the controller that he were beginning his takeoff roll. The captain interrupted the tail end of the copilot's readback with the comment "We're going."

The controller initially responded with "OK" (terminology that, although commonly used, is nonstandard), which reinforced the KLM crew's misinterpretation that they indeed had takeoff clearance. The controller's response of "OK" to the copilot's nonstandard statement that they were "now at takeoff" was likely due to his misinterpretation that they were in takeoff position and ready to begin the roll when takeoff clearance was received, but not actually in the process of taking off. He also most likely hadn't heard the captain's announcement that they were "going," since van Zanten had said it so soon after the copilot's readback. Aware of the possible misinterpretation, the controller then immediately added, "Stand by for takeoff, I will call you," indicating that he had never intended the clearance to be interpreted as a takeoff clearance.

However, a simultaneous radio call from the Pan Am crew at that precise moment caused mutual interference on the radio frequency, and all that was audible in the KLM cockpit was a heterodyne beat tone, making the crucial latter portion of the tower's response inaudible to the KLM pilots. The Pan Am crew's transmission, which was also critical, was reporting: "We're still taxiing down the runway, the Clipper 1736." This message was also blocked by the heterodyne and inaudible to the KLM crew.

Due to the fog, neither crew was able to see the other plane on the runway ahead of them. In addition, neither of the aircraft could be seen from the control tower, and the airport was not equipped with ground radar.

After the KLM plane had started its takeoff roll, the tower instructed the Pan Am crew to "report when runway clear." The crew replied: "OK, we'll report when we're clear." On hearing this, the KLM flight engineer expressed his concern about the Pan Am not being clear of the runway by asking the pilots, "Is he not clear, that Pan American?" However, the captain emphatically replied "oh, yes" and continued with the takeoff.

Further analysis revealed that the KLM captain's impatience for getting the flight under way (they were running late) and his intolerance for accepting critical feedback from subordinates were "signifiicant contributing factors" for the crash. Prior to takeoff, the transcripts made it clear that the captain did not appreciate being corrected by his copilot. At that time, a young pilot's career could be adversely affected by a senior captain, so they often choose silence even when they have significant safety concerns.

As a result of this incident, all airlines took a hard look at their cockpit communications and brought to an end the rigid adherence to status and rank that had dominated these exchanges. These were valuable lessons learned; unfortunately, it took 569 lost lives to learn them. The real tragedy is that this accident could have been averted several times over if just one person had had the courage to yell "stop!"

Can you recall times in your own career when the same would have been true—that if you or a coworker had yelled "stop!" a crisis would have been averted? To this end, I want to encourage you be a true industry maverick; embrace the establishment of high standards and by all means have the courage to live them!

11

Influencing versus Motivating

He [Bigfoot] had a near supernatural sense of exactly when, at what precise moment, one of his employees had had enough. He could tell when the bullying, the relentless sarcasm, the constant all encompassing vigilance had become too exhausting. When one of his people was fed up with staying awake all night anticipating his likes and dislikes, was sick of charting his mood swings, was tired of feeling demeaned and beaten down after being asked, for instance, to clean out the grease traps, was ready to burst into tears and quit, then suddenly, Bigfoot would appear with courtside seats for a playoff game, a restaurant warm-up jacket (given only to Most Honored Veterans) or a present for the wife or girlfriend— something thoughtful like a Movado watch. He always waited until the last possible second, when you were ready to shave your head, climb a tower and start gunning down strangers, when you were ready to strip off your clothes and run barking into the street, to scream to the world that you'd *never never never* again work for that manipulative Machiavellian psychopath. And he'd get you back on the team, often with a gesture as simple and inexpensive as a baseball cap or a t-shirt. The timing is what did it, that he *knew*. He *knew* just when to apply the well-timed pat on the back, the strangled difficult-for-him "Thank you for your good work" appreciation of your labors.

—**Anthony Bourdain,** *Kitchen Confidential*, 100

You can't motivate people. There, I said it. It's a bold statement, but I stand behind it. Motivating others is a myth. People come to the job site with their own intrinsic set of beliefs about such things as work ethic, personal responsibility, pride of ownership, and a general sense of what doing a good job means. Having said this, you may well ask: Does this mean that people can't change their outlook on life? Couldn't an irresponsible person learn to become responsible, and a slacker aspire to be a hard worker? Sure they can. In fact, we've all seen examples of just such turnarounds—maybe even in our own lives. But generally speaking, when this change occurs, it isn't due to something that a leader did. There is no magic button that will suddenly light the fire inside of someone if the spark doesn't already exist. What a leader can provide are clear standards, feedback regarding how someone's performance measures up against these standards, and his or her availability to provide assistance or suggestions for improvement. The rest is pretty much up to each individual worker. Trust me; no amount of inspired leadership can rouse a nonperformer unless it is already within him or her to be awakened. Understanding your limitations as a leader is actually a good thing. Many leaders assume too much personal responsibility when their employee's act badly and end up wasting far too much time, energy, and stomach acid, trying to figure out "how to get through" to them, often to their team's and their company's detriment. There is one such example that bothers me to this day. A project executive (PX) who had been promoted because of his sales ability proved most vexing to all who worked with him. He was intelligent, presented well, was good with people (at least on the surface), and at first blush, the owners loved him. But after the sale was secured, and the honeymoon was over, his behavior rapidly deteriorated. He disappeared for long periods of time and often failed to show up for meetings that he himself had called. Worse, he could never be counted on to produce any of his deliverables. The company spent close to two years trying to convince him to do his job; they praised him, scolded him, told him how much they needed him—but to no avail. In the meantime, his teammates put in extra hours picking up his slack and were often called upon to make excuses for his unexplained absences to an owner who was increasingly questioning the value he added to the job. All in all, a true Lean disaster. In the end, the company finally had to let him go. In retrospect, rather than being preoccupied with his potential, top managment would have been much better served cutting their losses before the PX's impact on productivity, quality, and owner perception became so detrimental. In the end, despite all of their good intentions, all they ended up really doing was enabling his Lean-killing behaviors.

While it's true that we can't motivate a person who isn't intrinsically motivated, unfortunately, we can *de-motivate* those who are. Demeaning sarcasm, public ridicule, obnoxious personal comments, and the distorted need to micromanage others are well-known culprits when team morale flags. But so too is the failure to provide clear expectations, meaningful recognition, and timely and thoughtful feedback. All of these things can whittle away at an otherwise effective performer's will and subsequent work product. So, what can we do to make sure that this doesn't happen? Though we can't motivate others we can, via our *influence*, either enhance performance or deflate it.

Here are the three primary ways that we influence people:

1. We can reward them for doing what we want them to do (positive reinforcement).
2. We can threaten to punish them if they don't do what we want them to do (negative reinforcement).
3. We can punish them for doing something we don't want them to do (punishment).

That's basically it. Everything else we do as a leader pretty much boils down to one of these three things.

To be an effective influencer, we need to stay mindful of the following:

- Have a clear vision of what the desired outcome should look like, better known as the goal or target.
- Have a clear idea of the behaviors and actions required to hit the target.
- Be engaged enough to provide meaningful feedback as to whether or not the behaviors, as they are being performed, will hit the target.

Get these elements right, and I guarantee you that you will be an extraordinary influencer. Get them wrong, and you'll have continuous problems with morale and performance—and a plethora of complaints to HR about your leadership style.

POSITIVE REINFORCEMENT

Positive reinforcement, simply put, is giving people a reward *after* they do something that we wanted them to do. The reason we should want to do so is that the research is unequivocal; positive reinforcement is the best way to increase the likelihood of getting more of the same behavior in the future. For instance, let's say that someone has put in extra effort to update the submittal log, and we make a point of giving him or her a public pat on the back for it. Provided that the person likes pats on the back, we should see a continuance of the behaviors it took to update the log in the future. Further, if witnessed by others, this should serve to encourage them to step up their game in anticipation of similar rewards.

Sometimes leaders object to the use of positive reinforcement on the grounds that it appears manipulative. At first blush, I can understand the assertion. But manipulation connotes pure personal gain. That's not what is really going on here. While leaders may indeed benefit by someone doing their job well, what they are really providing via positive reinforcement is enough clarity about what is desired to allow someone to be successful in his or her own right. The benefit derived is mutual.

To be an effective positive influencer, you need to keep a few important elements in mind:

Rewards always follow the targeted behavior: The most common error people make when attempting to provide positive reinforcement is giving the reward too soon. A reward must always *follow* the desired behavior. If the positive reinforcement comes before the desired behavior, this is called a bribe, which may or may not be effective. For example, if the desired goal is zero lost time accidents for one month, but up front, all the workers are given an all-in-one tool (the intended reward), there is no longer any additional incentive for them to act safely because they already have the reward!

This is the inherent problem of rewarding business development people with bonuses that are contingent only on sales made. If the bonus isn't directly tied to a job's long-term profitability, what's the incentive for the salesman to land good work versus bad?

Consistency: The target and the behaviors required to hit the target need to be predictable versus constantly changing. Moving targets drive people insane. Sometimes we become inadvertently inconsistent because things are moving so quickly, or we receive such conflicting feedback from A&Es and owners about what they want, that the subsequent target we give to our people keeps shifting. This is an easy enough situation to rectify, by simply slowing down and informing the team of the problems you are having in obtaining a consistent message from your external partners. As long as you keep them informed, most teams will remain flexible and cut you some slack about the confusion generated. But there are a small number of managers who think that keeping their people in the dark and maintaining a level of unpredictability is a good thing—and do so intentionally. They think this "keeps everyone on their toes." But the only thing this behavior is good for is maintaining a weak leader's power base by appearing to be the only one who is always most on top of things. If you are one of these misguided souls, do your team a favor: change your viewpoint. You are not doing them, or in the long run, yourself any favors by building your personal power at the expense of your team's productivity (and sanity). A lack of consistency creates team disharmony in the worst way possible: it makes everyone hesitate because they have to guess at what is right. People need to know what the target is and the activities and behaviors that are consistently required to hit the target. If you truly want to be able to delegate some of your tasks off of your plate, rather than just complain about how overloaded you are, you will need to let people in on the plan. But if you are addicted to being seen as the "hero," keep doing what you are doing until you burn out or have a heart attack.

Timing: To get the biggest big bang for your buck, positive reinforcement can't be put off for unrealistically long periods of time. There is a distinction between celebrating a goal or milestone and reinforcing the behaviors that will get you there. Waiting two years or even two months to receive a reward loses any of its instructive or incentive punch; it is too far in the future to carry any meaningful weight in the present and will be forgotten in the day to day. But a reward that is immediate draws attention to what is desired right away. Giving a pat on the back for a well-vetted purchase order (PO) is instructive. Waiting to do so two months down the line after a milestone is accomplished is not.

A reward must be realistic: Similarly, if the goal is completely unrealistic to attain, let's say three years of zero lost time accidents, then people will see you for what you truly are—not safety conscious, but incredibly miserly. If you truly care about the goal, then you need to break it down into its constituent parts (consistently wearing protective gear, keeping work areas clean, etc.) and intermittently reward the behaviors that will produce the targeted goal.

The reward has to be rewarding: A reward, by definition, is something that a person wants. Otherwise, it is not considered reinforcing. A corrugated box plant learned this lesson the hard way. They wanted to improve their attendance numbers, so they decided to have a competition (rarely a good idea) to reward perfect attendance over a six-month period. So they posted graphs and charts, and as others fell away, two people emerged as the final contenders. On the last day of the competition, both people called in sick. Why? It had a little something to do with what the company picked as a "reward." The positive reinforcement selected was dinner out with the general manager and his wife at a swanky restaurant. These were hardworking, working-class folks who didn't consider going out to dinner with the boss a reward. They viewed it as an uncomfortable, possibly humiliating evening. So they bailed. They would rather have received a company jacket or a supermarket gift certificate than a night out where they would have had to dress and act in ways that were uncomfortable to them.

To find out what is actually rewarding to your people you need to be engaged with them enough to know what actually fires them up. Teams vary greatly in this regard. Some want social events (dinners, company BBQs where they can bring their families), while others appreciate something tangible, like bonus checks or company apparel. A group of plumbers in southern California worked their guts out in exchange for the promise that their general foreman would allow his head to be shaved in the company parking lot if they were able to complete their portion of the work ahead of schedule.

Make it a daily practice: Managers often fail to see how a kind word or gesture can make all the difference in the world. The simple act of walking around and catching people doing something right (as opposed to catching them doing something wrong) goes a long way to boost people's morale. It also helps to build in quality. Put yourselves in your employee's shoes for a moment. If you do something right and are caught in the act and positively called out on it, wouldn't you be far more likely to repeat the same actions? Do enough of these right actions consistently, and the likelihood of you producing a high-quality product is much greater.

Conversely, if these same behaviors go unrecognized for prolonged periods of time, they will slip. Failing to give any meaningful recognition or praise for the behaviors that we want unintentionally puts them on, what we call in the trade, an extinction curve. By not rewarding the behavior, it slowly goes away. Using slot machines as an example, a person will remain very motivated for the first twenty or so pulls as the anticipation of big payoffs dances in his or her head. But if he or she goes beyond twenty pulls without being rewarded, he or she will either move on to another machine or quit playing entirely. That's why casinos program their machines to pay off in small amounts at unpredictable intervals—it gives people just enough incentive to keep at it! The same principle is true at work. Think way back to that job where you toiled away cranking out high-quality submittals, one after the other, and your boss never said a word. Toward the middle of the job, didn't your motivation flag? It probably only picked up again when the end was in sight and the hope of moving on to a better opportunity came along. So, why replicate such a soul-deadening environment?

Similarly, if we give information and training only to those we like, and not to those we don't, those who we have chosen not to give our time and energy to will also have been placed on an extinction curve and their performance will predictably suffer—thus seeming to prove our ill feelings toward them. Never forget that receiving necessary information is in itself reinforcing, as it increases a person's ability to be successful. Therefore, withholding it feels like a punishment, which is why many managers feel justified in doing so when they are angry or upset with someone, and why employees are so angered when this is withheld.

Please don't assume that a paycheck is a positive reinforcement. Paychecks are only indirectly tied to performance; you get one whether you are a great performer or just an okay one. Paychecks do very little in terms of influencing performance—even in a bad economy.

Rewards don't have to be expensive—or even tangible: You don't have to break the bank to reward people. A pat on the back, public recognition, or a personalized handwritten or emailed note of thanks for a job well done will go a long way to help sustain quality performance—provided that it is done in a sincere and meaningful way (specific versus vague).

Positive reinforcement is the most powerful tool that a leader has in his or her tool belt. It is the type of influence that people respond to most favorably. It impels them to *want* to do the behaviors that we'd like them to engage in. Again, if you doubt the power of positive reinforcement, take a return trip to Las Vegas. Periodic payoffs at slot machine and other gambling activities are classic examples of the power of intermittent positive reinforcement. They induce people to stay up until the wee hours engaging in the same behaviors over and over again. The fact that a new $1 billon hotel/casino can be paid off in about three months testifies to the power of this principle!

NEGATIVE REINFORCEMENT (THREATS OF PUNISHMENT)

Negative reinforcement is a contingency whereby someone can prevent a punishment from happening by doing something that we want him or her to do. It is time tested, it works, and it has been used to rule our daily lives from time in memoriam. "I won't burn your village if you give me all your gold," "I won't kill your family if you consent to marry me," "I won't banish you from the kingdom as long as you keep telling me how wonderful I am!"—these are just some of the timeless classics from the negative reinforcement hall of fame. Today, negative reinforcement contingencies are a bit more benign, but just as powerful. "I won't give you a speeding ticket as long as you follow the speed limit," "We won't fine you or throw you in jail as long as you pay your taxes," "I won't holler at you provided that you come home from work on time"—these are just some of the many things that dictate how we will choose to behave in our daily lives. Excluding masochists, would anyone submit to having sharp metal instruments inserted between their teeth and gums if it didn't hold the promise of preventing gum disease and cavities in the future?

On the job, negative reinforcement is an everyday occurrence. We complete schedule updates so that the owner won't be upset with us. We submit our field payroll reports and billings on time so people won't scream at us for not getting paid. We maintain our plan room so the inspector won't march off in a huff and say bad things about us. The truth is, negative reinforcement flat out works. But there is a downside to its use; as powerful of an influencer as it is, most people typically resent it when their behavior is controlled in this manner. In other words, we will comply, but we certainly won't like it—and we'll remember being treated in this way for a long time. If you are the kind of leader that relies on threatening to breathe fire on your team should they mess up, you should know that you will get the behaviors that you seek, but you'd better plan on being on the job 24/7. Because the research is very clear: once the threat of punishment (you) is removed from the equation, your people will ease up and reduce their performance to prethreat levels (better known as the "phew, he's gone" factor). Also, under the threat of punishment, people will only engage in enough of the behavior to prevent the punishment from happening—nothing more. Cool Hand Luke may have shaken the bush, but he certainly wasn't about to do anything more than that. In Lean terms, the assembly line will continue to move, but only as long as you are there to keep it moving, and only at the rate that keeps the punishment at bay. Under this contingency, people engage in behaviors not because they want to, but because they are afraid of what will happen to them if they don't. Because of this, negative reinforcement isn't effective for helping people to learn new behaviors. In such an environment, learning something new just means that there is one more thing for the boss to hold over their head like the sword of Damocles, so why bother?

PUNISHMENT

Punishment is effective for one thing and one thing only—to stop unwanted behavior from happening. "I'm writing you up for being late. One more write-up and you're fired!" "I'm chewing you out right now because you failed to get owner approval before you did that additional excavation work." By punishing the person (delivering something unpleasant that they don't want), we are attempting to stop them from doing something that we don't want them to do. But be very clear; in most instances, punishment is only part of the equation. Yelling at someone for not getting owner approval only tells that person what he or she should stop doing; it does not instruct him or her on how to prevent these types of situations from occurring in the future (such as finding the correct passage in the contract, reading it, understanding it, seeking out the owner's rep, writing an approved letter etc.). If we rely on punishment alone to correct improper performance, we are essentially leaving it up to the other person to figure out how to achieve the desired behavior, which may or may not happen. That's why after delivering a punishment it is important for a Lean leader to spend time helping the person figure out what they should be doing differently in order to perform up to the established standard.

Having said this, there are certain behaviors that don't warrant a second-chance "bite of the apple." Stealing, lying, and willfully misrepresenting work product, in my mind, are always fireable offenses that don't merit our coaching time. It's not your job to help people become less injurious sociopaths. These are not simple correctable mistakes; they are issues of character and warrant punishment.

Rewards, punishments, and threats of punishment are extremely powerful tools if used correctly. If you think back to the bosses that you've had that you considered effective, like Bigfoot, I'll bet they were all extremely good at knowing exactly when to give you a pat on the back or a swift kick in the butt!

SHAPING: HOW PEOPLE ACQUIRE NEW SKILLS

Let's say you want someone to acquire a new skill, for instance, writing a master schedule for the first time. Would you say, "Just do it," and leave him to his own devices? If so, let me suggest an alternative.

The first thing to do is to establish a clear vision for the targeted behavior. To do this, you would show the person a master schedule from a similar type of project, review its features, and point out some specific elements to incorporate. In other words, you'd provide a clear picture of what is to be modeled so that the person doing it for the first time can approximate it. In essence, what you are saying is: "Here's the template; I want you to produce something very similar to this."

Next, you would begin to show him the steps required to make the new schedule happen. You could have the person look over your shoulder as you break out the first couple of weeks of activities—just so he can see, mechanically, how to go about it. Encouraging the person to ask questions at this point will stimulate his engagement and allow you to fine-tune the message.

After this, it is important for you to coach him on the specific ways to interact with subcontractors in order to illicit their input, feedback, and buy-in (i.e., meeting with subcontractors to review *their* schedule of activities). It's usually a good idea to have the person observe you run this type of meeting first so he can get a sense of what the expectations are. Next, have him run a meeting in your presence, so afterward, you can give him feedback on what he did and didn't do so well and help him to incorporate the information gathered in the meeting into the new schedule.

Once you are reasonably assured that the person can work independently, then you can turn him loose. But when you do, make sure to build in dependable times for check-ins and Q&A to make sure that he is still on track.

At each step along the way it is vital that you provide praise when the person hits the target and corrective feedback when he misses. This is how to literally shape someone's behavior toward a targeted goal.

Now, let's say that the person you are coaching is particularly shy and is avoiding calling the subcontractors. How should you handle this? First, it is important to understand that this is a common fear and yelling never helps anyone overcome it. Second, reassure him that everyone goes through these trepidations when they are in the early stages of their career. But if this empathetic stance doesn't help and he digs in his heals, you'll need to let him know that obtaining subcontractor input is not optional and that continued avoidance to do so is unacceptable. The use of the threat of punishment here is not to harm, but to break down the person's natural fears of engaging in a new and unfamiliar set of behaviors. It is important to ask him, at this point, what further help he may need from you in order to get started. Once the person starts making phone calls to subcontractors and follows through with the required meetings, praise him for engaging in these new and uncomfortable behaviors.

This same formula holds true for any new process that you are trying to teach; show relevant high-quality examples, let people watch you do one, let them do one, and then provide immediate feedback and coaching as to how they did. Once this has been established, turn them loose and conduct intermittent check-ins and coaching as needed, while making sure to praise behaviors that hit the mark.

This takes more up-front thought and effort than you may be used to, but in the long run, it will prove to be a time saver. Shaping is the most reliable way to help people to acquire new skills. If you can master this formula, you'll also reduce your need to micromanage (fear that people will do things wrong) and you'll do your part in both keeping the line moving and building competent employees for the future.

TEAM REINFORCEMENTS

Too often, when we think of providing reinforcements, we think only in terms of individuals, and miss the opportunity to use this same tool with our entire team.

Garner Gremillion, PX currently with Bovis Construction, is a master at delivering positive reinforcement at the team level. On the P2D4 job (a fast-track Intel research facility with a clean room and unique tool install requirements) he laid out a series of three-month milestones for the team to hit, and each week gave the team feedback (in graphic form) as to how they were progressing. Further, he outlined what would need to happen the following week to maintain momentum, and invited people to give input, express concerns, or ask for help in order to stay on track. He then made a point of checking in with people throughout the week to give praise or corrective feedback as required, and was careful to ask what they needed from the management team to help clear any obstacles that may have arisen along the way. And when the team accomplished the designated milestones, he celebrated them like crazy!

Over the course of a twenty-eight-month project, the team missed just one milestone; given the fact that they were putting a whopping $32 million of work in place per month, I think you'd agree that Garner's approach was pretty effective.

HOW THINGS CAN GO ASTRAY

All of this sounds pretty straightforward, doesn't it? But things can easily go astray when managers confuse the contingencies by either inadvertently punishing people for doing something good or rewarding them for doing something bad. Let me give you a few examples.

A project manager constantly complained about the performance of a young subordinate. "She never looks ahead. She'll sit there twiddling her thumbs until I dole out the work, and she never does anything more than I assign her." Fair enough, I thought—one of those generation Y things. Except that when I talked to the young lady in question, and her teammates, a far different picture emerged. According to them, the project manager (PM) was the poster boy for Control Freaks Weekly. If anyone on the team did anything that he hadn't personally directed them to do, he squashed them like bugs—regardless of the outcome of their actions. In his mind, things had to be done perfectly, which meant doing things in the exact same way that he would have done them. Can you identify the reinforcement contingency that the PM inadvertently set up? In short order,

people learned to wait for him to give assignments and specific instructions, rather than initiating anything on their own. He was actually reinforcing people for waiting for assignments and punishing their initiative.
His people also learned that they could prevent a punishment from happening (negative reinforcement) if they only did what they were told—and
nothing more. Contrary to his belief, the PM was actually instrumental
in creating the very thing that he was complaining about so vehemently!
The devastating ramifications for developing a Lean culture in such an
environment should be apparent. The focus on continuous improvement
goes out the window when all anyone on the team thinks about is how to
avoid punishments.

Unfortunately, controlling behaviors such as these are very resistant
to change. Every time a control freak lets go of something, and someone
drops the ball (a natural occurrence when people are learning new skills),
it reinforces in his mind that he should *never ever* delegate anything to
anyone else ever again, and so he doesn't.

Unintended reinforcement contingencies crop up all the time in large-
scale endeavors, such as when a company decides to retool its processes
or procedures. For example, ten years ago, in an attempt to bring greater
standardization to project cost reporting, and to minimize the occurrence
of errors, a general contractor installed Prolog as their official project
management operating system. They spent millions on physical implementation and training their people in its use. But there was an inherent
problem: people were uncomfortable using it, as it was laborious to set up
and tended to slow them down. Besides, they were used to customizing
their own Excel spreadsheets and, in the heat of the moment, were highly
resistant to learning a new way of doing things. In the name of expediency,
PMs and PXs often looked the other way when people bypassed Prolog
and continued to use their own spreadsheets, provided that, in the end,
they got the product that they wanted. (This was exacerbated by the fact
that a lot of PMs weren't at all comfortable using Prolog themselves.) What
was the long and short of this looking the other way? It meant that their
cost department continued to get wildly varying budget reports from each
project, which slowed down their productivity and increased the possibility of errors—the very problems that the company was trying to avert
when it installed Prolog in the first place! Additionally problematic was
the fact that the overall implementation of Prolog was artificially delayed.
This meant that the company had to spend additional time and money

retraining people on something that they had already been trained on because they had forgotten what they had learned while they continued to use the old systems. Looking the other way not only inadvertently reinforced people to keep doing things the old way, but it put newly learned behaviors on an extinction curve. This is waste with a capital W!

Here's another example. A GC on an airport job in California was struggling mightily. For some reason, the team simply wasn't able to gain any traction. Even after getting over the hump of an unanticipated contaminated ground issue, progress in the field was slow. Buy-outs were also happening at a snail's pace, and this led to additional delays. The managers were beside themselves with worry and, as a result, were putting in tons of additional hours. Meanwhile, the rest of the staff seemed blissfully oblivious. They would put in their eight hours, go home, and unlike the managers, didn't seem to be carrying the weight on the world on their shoulders. As the project continued to lose ground, the owner let it be known, at a national port authority gathering no less, that they were none too pleased about the GC's progress, thus jeopardizing future work around the country. (Yes, it is indeed a very, very small world.)

When I met with the managers and pointed out the difference between their demeanor and that of their staff's, they had plenty of theories to account for it.

"You know this generation today; they just don't want to work that hard, and let's face it, they simply don't care in the same way as we do," said one. (There were murmurs of agreement.)

"I think it's because we don't hold the staff accountable enough," said another. "We let them slide on things we shouldn't." (More murmurs, and a few harrumphs.)

"I don't think people fully understand their jobs," said another. "I think we have a lot of people that just don't get it. I think we need to do a better job of helping people understand their roles." (More murmurs, but not quite as many harrumphs.)

Then I asked another question: "How would you describe the execution and overarching philosophies on this job? Is it to please the customer at all costs? Keep going until the owner tells us to stop? How would you describe it?"

One manager's response was emphatic: "We bend over backwards to please this customer! We are constantly running things past them and looking for their feedback!"

"And how is that working out for you?" I asked.

He didn't answer. He just glared at me.

Suddenly, the PM jumped out of his chair as if he'd been hit with 120 volts. "Do you know what? The people out there are doing exactly what we trained them to do! We keep telling them to wait and make sure that they check in with the owner and get their feedback and approval before they do anything. Instead of building what's on the drawings, they are waiting for the next addendum to come down, which, with this owner, can take forever!"

I said absolutely nothing for the rest of the meeting. I didn't have to; the managers had their answer. In a subsequent meeting they announced a change in philosophy to the staff—and the subsequent change in reinforcement contingencies.

"People, from this point on, we're building what is on the drawings. You let me worry about changes and interfacing with the owner. From now on we're moving forward until they tell us to stop. We need to stop waiting and start doing!"

What was the staff's reaction to this change in philosophy, you ask? One word: finally! As it turns out, they were sick and tired of the way things were going, and never fully understood why they were always being told to rein in their activities. But rather than buck the constraints, they quietly acquiesced to the judgment of their managers, who they assumed knew better.

I don't want give the impression that productivity on the job changed overnight, or that there weren't some significant issues with some personnel on the job. But I can say that once the managers changed the reinforcement contingencies (rewarding action versus inadvertently rewarding passivity), the project moved forward at a much more effective rate. Even better, this renewed vigor compelled the owner to award the company another, much larger phase a year later.

To this end, please don't underestimate the importance of helping the owner to change their own reinforcement contingencies around. In the above example, the GC kept bending over backwards to accommodate what the owner thought was important to the end users (the airlines), that is, flexibility around changes in design. But the more the GC did this, the more the end users became upset. The reason? Accommodating late changes was only an apparent target—not the true overarching philosophy of the job. More than anything else, the airline wanted to be able to operate out of the new gates ASAP. It became the GC's job to help the owner see the realities of what was truly important. Accommodating an endless stream of late changes was actually killing the most important goal (early completion) rather than helping it!

So the next time you are in a situation where you are not getting what you want, take a step back and be objective about what you have been rewarding and punishing. You may just find that you've hinked up your own Lean process by rewarding and punishing the wrong things.

12

Constructive Discipline (Knowing Where to Draw the Line)

> Bigfoot understood—as I came to understand—that character is far more important than skills or employment history ... He understood ... that a guy who shows up everyday on time, never calls in sick, and does what he says he's going to do is less likely to fuck you in the end than the guy who has an incredible resume but is less reliable ... Skills can be taught. Character you either have or you don't.
>
> **Anthony Bourdain,** *Kitchen Confidential*, 96

This business is filled with characters, and is one of the reasons that it is such a joy to be a part of. But unfortunately, there will come a time when someone's performance, behavior, or attitude is so out of variance with what is acceptable that you will need to draw a line in the sand. For the sake of the team, and your company, you will need to set a firm limit, which may include termination. Though you may have read that several Lean manufacturing enterprises guarantee their employees employment for life, you should know that this doesn't mean that they give people carte blanche to do whatever they like. They extend this guarantee in exchange for a promise: that the employees will remain devoted to serving the company's customers and its continuous improvement process. To that end, it is essential that you address those who act in opposition to these same principles in no uncertain terms.

Most often, when we think of taking corrective action, we do so with a specific individual in mind. And indeed, this is the most common occurrence. But when attempting to build a Lean culture, we sometimes must take action with a broader context in mind.

When Tom Sorley, CEO of Rosendin Electric, was named president in 1993, he inherited a company with a reputation of being "claim's artists" as a result of numerous internal problems. But with the help of his senior management team, they have turned Rosendin into the contractor of choice for most general contractors, not only in the San Francisco Bay Area, but in other areas where they have ventured as well. Tom is the epitome of a "Texas gentleman"—unassuming and approachable to a fault; yet he pushes for high standards, not by the whip, but by setting firm limits and investing copious amounts of his time with people. I happened to be sitting with Tom when an irate division manager marched into his office to complain about the poor service exhibited by one of the accounts payable (A/P) representatives for failing to collect the proper amount on an overdue billing. Unfortunately, blaming others—usually of lower stature—had become endemic at Rosendin over the years, particularly when failures of a financial nature occurred. Nonetheless, Tom listened attentively, and then with the utmost sincerity, asked the division manager what he thought an acceptable error rate for an A/P person would be. Caught off guard, the division manager blurted out, "I don't know—5%!" Tom reached into his desk and pulled out data that showed that the actual error rate for the entire A/P department, on average, was well below 2.5%. Then, without raising his voice, and while demonstrating the utmost regard, he continued, "I know that you want to get paid in full, and I would expect nothing less from you, but let me ask you something; who knows this job better, the A/P person who lives in a little cubicle way up on the third floor, or you and the PM under you, who is on that job every single day?" When the division manager acknowledged the obvious, Tom drove the point home. "So, why are you guys leaving it up to a clerk to sort out the intricate details of a job that she couldn't possibly know the answers to? Isn't that the management's job to handle?" After the division manager sheepishly agreed, Tom finished the discussion by making a plan for follow-up that left the division manager's dignity intact, but pushed the need for culture change. "So, after you and the PM get together and figure out how to go after what we are owed, I want you to come back and let me know if there is any help that you need from me—provided that it doesn't include handing things off to A/P."

This was a discussion that had to be repeated in various forms with numerous middle managers, but it was through such limit setting that Tom and the top management team were able to transform Rosendin's culture. Even in these rough economic times, I don't know anyone who doesn't have an undying loyalty and respect for what Tom, Larry Beltramo, and Jim Hawk have done in terms of standing up for what is right, good, and Lean at Rosendin.

Here is another example. Paul Pettersen, a retired VP and operations manager (OM) for Turner Construction, despite being a bit of a fire-breather, had the uncanny ability to get a team back on track when they were threatening to come off the rails. Despite the fact that I often felt like a neurotic poodle in his presence, I had and still have tremendous admiration for what Paul could do. You always knew where you stood with Paul. No one ever had to wonder what he was saying behind their backs, because he never had a problem saying it to their face. He also had the ability to walk a job site, and within five minutes, tell you exactly what was being executed properly and what wasn't—and he was seldom wrong. On one such occasion, after walking the job and seeing a high degree of disorganization, he asked to attend the afternoon staff meeting. After fifteen minutes of witnessing the on-site managers talk among themselves as if the rest of the staff were invisible, he could take no more. He dismissed all but the managers, and once they were out of earshot, he let the managers have it. "Just make a decision!" he exclaimed. "At this point, I don't care if it's the wrong one, just make a fucking decision!" Stunned, the managers looked at him like deer in the headlights. Seeing their terror, Paul backed off a bit.

"Look, I heard all of your ideas, and none of them are bad. But what is killing this job is indecision. Instead of waiting for the perfect one to emerge, pick a direction and go with it until the data says that you need to go another way." For the management team, this was exactly what they needed to hear. They were so caught up in building a consensus among themselves and so worried about not making a mistake that they had tied themselves and their team up in knots. This confrontation helped to dislodge them from their self-inflicted mire and move forward.

To help clarify your thinking about whether or not you need to engage constructive disciple on a broader level, and as a bridge to building a more committed and responsible organization, please keep the following warning signs in mind:

- Continually having to drop down to do someone else's job means that someone hasn't done his or her job properly. This is detrimental, because who is going to do your job as the orchestra leader if you continue to have to drop down to do someone else's job?
- If you have to drop down to get things done properly, you likely have an unqualified person in that position. A top leadership's job *is not* to do other people's work for them, but to relentlessly focus on

getting the right people into the right positions so drop-downs are not necessary.

- Leaders don't make assumptions or absent themselves from responsibility; they ask the extra question and make sure that their areas of responsibility are always covered. They never give themselves permission to drop a deadline. If you have someone under you who is in a leadership position who frequently makes assumptions or absents himself from responsibility, replace him.
- Leaders are not victims. They step up and take ownership rather than making excuses or blaming failures on others. If you have someone under you in a leadership position who continually whines or points the finger at others, replace him.

In the examples cited above, the people involved were competent and of good character; they just needed to be jolted out of their mindset. But let's take a look at some examples that are more dubious and nefarious in nature.

There was the project manager who, at a company function, didn't quite get that his sexual advances weren't being reciprocated by a young female subordinate. So after a couple of more drinks, he went back to the trailer and decided to make them even clearer—in writing, via time-stamped email—so there would be no ambiguity for the young lady in question—or her attorney.

There was the hyperintelligent OM who, at the drop of the hat, would launch into heavily opinionated political discussions and hyperbole. On one memorable evening, he decided to take everyone out to the bar, including some architects, got inebriated, then proceeded to jokingly complain about how "the faggots" were taking over the country. Catching the uncomfortable expressions of the architects seated at the adjoining table, he corrected himself and assured them that he considered them to be "good faggots." Did I mention that the city he worked in was San Francisco?

Then there was the VP/general manager who regularly sidled up to his secretary as she performed some last-minute work and threatened to fire her if she wasn't done by the time he left the office for the day. When she had finally had enough and lodged a complaint to HR, he claimed that he had merely been joking.

Then there was the diversity manager who wouldn't let his underutilized administrative assistant help anyone else out on the project, even though they were swamped, because he didn't want to let it be known that he and his administrative assistant didn't have all that much to do.

Then there were the project engineer and project superintendent who hated each other so much that the only way they communicated was through vitriolic emails that they copied to everyone else on the team. The emails were so bile filled, and created such an us versus them mentality, that when other people rotated onto the project to close it out, nobody could figure out what either side of the house had done to track the job, thus causing delays that cost their company $2 million of potential bonuses for early completion.

Then there was the superintendent who complained ad nauseam about all the extra hours he had to work because of the incompetence of his teammates. He did, in fact, log a lot of overtime. Unfortunately, a quick scan of his computer usage revealed that he spent 75% of it "working" on NakedLatinaPics.com.

There was the infamous VP who, on frequent occasions, directed his subordinates to take very specific actions, and would gobble up all the credit if the outcome was good, and feigned a complete lack of knowledge if it went bad. He was even known to cheat on his golf scores at company functions.

There was the superintendent who complained bitterly about being ill-used by his project director (PD) because the PD dressed him down publicly after a wall, which was improperly clipped, collapsed, nearly killing several workers. Instead of frantically checking to see if the other three walls under his direction were similarly flawed (they were), he instead chose to march over to human resources to file a complaint against the PD.

Then there was the executive who made a point of parading around his new girlfriend at company functions. He even went so far as to post pictures of himself and the young lady on the company Web site—much to his wife's and his CEO's dismay.

There was the PM who took every opportunity to bash his project executive (PX) behind his back. And when any underling dared to speak up and suggest that the PX actually had some good ideas, the PM accused the person of being a traitor and subsequently withheld critical information from him or her and went out of his way to sabotage the underling's work.

Last but not least, there was a superintendent who hated Intel with such zeal that he made a career out of working on their projects. When I saw his anguished face, I couldn't help but ask why he intentionally put himself through so much pain. "Don't you see?" he said. "I know them so well that I know how to screw them! It's all I live for—to stick it to them each and every day!" (I have to admit that I have mixed feelings about this one. While Lean is all about the customer, anybody who has spent any appreciable time working with Intel, and been subjected to their needless and senseless instigation of negative conflict, and *doesn't* feel the urge to fill a pillowcase with bits of scrap rebar and beat them senseless with it, well, they just aren't right, if you know what I mean.)

Fortunately, in my fourteen years of doing this work, I have to say that the number of truly bad "bad actors" that I have encountered has been miniscule compared to those who anonymously toil away each day to do their very best. Nonetheless, the above examples are important as they pertain to Lean culture. They were not included for shock value or to increase book sales. Incidents of sexual harassment, offensive behavior, and abuses of power serve as impediments for smooth flow, continuous improvement, and a client-centered, value-added approach by forcing people to think about something else entirely—and that something is *doubt*. Such ugly behaviors stir up doubt about a company's true philosophy, doubt about the integrity of its leadership, and doubt as to whether or not to remain an employee with the company. That's why such failures of character that occur can have such devastating effects on morale, and why they must be addressed in a decidedly meaningful way—because they cast a pall on the entire company and everyone in it.

Don't get me wrong; I love the fact that there are characters in this industry. In a world that is far too homogenized and sterile, where few people speak their minds, and even fewer mean what they say, it is refreshing that there are so many people in construction that are willing to do both. And personally, I have no problem accommodating a wide range of behaviors—even the aggressive, the strange, and those deemed politically incorrect—as long as in the process of what they are trying to accomplish, the leaders are doing so for the greater good. The thing I ask myself is this: Was the person trying to do what was best for the project and the company, or was he merely doing it to benefit himself? I know that this seems somewhat subjective. But honestly, in my mind, it's not that difficult to discern. If someone meant to do good, but couldn't get out of his own way because of a quirk or skill deficit (i.e., how to express his anger in ways that are productive), I'll work with him until the cows come home.

But for those who are clearly selfish, self-centered, or otherwise out for themselves—who lie, steal company time by not doing their job, ruin the company's hard-earned reputation through careless acts, aren't interested in giving their best, or worse, who abuse their power by tormenting those below them for their own personal gain—well, I say, "Adios, see ya later, ta-ta, have a nice day, and don't let the door hit you on the backside on the way out!" I have no patience for them and I recommend that you don't either.

On occasion, I've taken some heat for refusing to work with some individuals. Since they often make money for their company, many executives reach out to me for assistance rather than pulling the trigger and firing them as they know they should. But my reasoning is simple: I'm not interested in helping sociopaths become more skilled at their craft. If someone only has his own interest at heart, and has no qualms about harming others in the process, and there is no objective reason to account for the abhorrent behavior, then I say, to quote Tony Soprano, "Forget about it!"

The same goes for those not in leadership positions. To paraphrase Anthony Bourdain again,

> … I don't care if this person was, or is, your best friend. If his values stink, if he bad mouths you and the company at every turn, if you can't find him when you need him, or doesn't pull his weight, and could care less about making the other people on the team that he works with look good—can him! In a couple of years, when he's moved on and (hopefully) matured—he may just call you and apologize for all the crap he pulled.
>
> —**Anthony Bourdain**, *A Cook's Tour*, 236

I couldn't agree more. This may sound harsh, but being hard when you need to be is important. There is no bigger morale killer on a team than to witness a coworker act badly and see nothing done about it. As a manager, you might as well stamp the person's behavior with a big seal of approval, because by choosing to ignore the problem, that is exactly what you are doing. In essence, you are reinforcing it.

Let's now turn our attention to a third category. What about those who are of good character, but are clueless in terms of their team-damaging performance? Feeling sorry for them or ignoring their deficits won't work. And we know how deadly and detrimental dropping down and doing their work for them is. So what's the answer? How do we deal with these people?

Here are some guidelines that you can use to help correct the poor performance or inappropriate behavior of someone who is skilled, but unaware of the impact that his or her negative behavior is having on others.

Operationalizing problematic behaviors is a fancy way of saying, "Describe the problem behaviors in observable and measurable terms." This is important, because if you are going to try to help someone change his or her behavior, you will need to know what the preferred outcome should look like, and how far the person in question's performance deviates from the standard.

Let's take a situation that might appear vague at first and translate it into operational terms. (The example we will use is a bad attitude exhibited by the project receptionist.)

1. What are the observable behaviors that are problematic?

   Answers the phone in a terse or rude manner; is short with people; hangs up abruptly without the appropriate salutation; often is sarcastic when someone comes into the trailer and is unsure of who he or she needs to speak to.

2. Why is this behavior a problem?

   Sends an arrogant "we don't care" message to clients and the people we work with. Customers have actually commented on and complained about the behavior.

3. What behavior would you like the employee to substitute for the current behavior?

   Our customers and the people that we work with are our life's blood. Be courteous. Show interest and concern. Treat people with respect. Act "as if," that is, talk to each person as if they were someone in your life whom you cared about and respected.

4. When do they need to demonstrate that they have corrected the problem?

   Right away.

5. What will happen if the problem is not corrected?

   Termination.

6. What is your schedule for follow-up?

   Feedback will be provided at the end of each day to note improvement or what still needs to improve. A formal review of performance will be provided once a week for one month to discuss what further actions might be required. If, after one month, the behaviors are in line with expectations, we will formally review performance each quarter.

When presenting a personal improvement plan such as this, I recommend creating a formalized, witnessed, written plan. When you present it, make sure to do so in private (with an HR representative present). Also, make sure you present it in such a way that the person believes that you actually want him or her to succeed at it. And don't forget about this next part: As soon as the person starts to exhibit the behaviors you are looking for, praise him or her for it—this is the other part of the equation that makes disciplinary actions successful. But if he or she should fail to demonstrate appreciable improvement, or further complaints are lodged, terminate for cause.

This last part is essential: if you do need to initiate such a plan, please, please, please don't do so in a vacuum. Inform your boss, and seek out an HR representative for assistance, guidance, and help with the proper documentation guidelines!

13

Commitment and Accountability

Though the themes of commitment and accountability have been woven throughout the fabric of this book, it is important that we highlight some particular areas where leaders can exert their influence to fully promote these key elements.

To truly call an involuntarily assigned group of individuals a team, there needs to be a sense that each person belongs to the larger whole. In addition, everyone on the team must be fully cognizant of the negative impacts he or she can and will have on his or her teammates should he or she fail to execute his or her responsibilities in a timely, high-quality fashion. As Tarpey, Konchar, and Grinnell eloquently describe in their article entitled "Forging a Leadership Culture":

> People, with their tendencies, strengths, weaknesses and general disposition combine to create a culture within a project team. The team is comprised of a variety of individual personalities, trained in a unique discipline each offering a unique set of experiences to the team. The ability for these individuals to join together and create an environment that promotes timely, accurate and useful communication of data or flow of information is nested in the dynamics of this team. Most failures which occur on projects can be traced back to a breakdown in communication, an unfulfilled commitment or a lack of information delivered from one team member to another. Because of the extreme interdependency of tasks that are organized to deliver a project, a domino effect is introduced when commitments are broken or critical pieces of information are withheld. Our focus therefore, should be on the development of people who are trained to first properly manage themselves and then to manage the network of commitments that are developed on a project.

Did you notice the critical punch line in this paragraph, that is, the notion of self-management being the key to effective teamwork? But what is meant by this, and how is it developed if it is found lacking?

The ability to self-manage is paramount if a team is to consider itself committed and accountable. Embedded in this concept is a deep awareness that others on the team are counting on us to manage our time effectively so that their portion of the work can be delivered to the line on time and of the highest quality. Failure to properly self-manage will necessarily generate waste and negatively impact every teammate up and down the line. Yet as we all know, at some point, someone on the team will fail to live up to his or her expectations. So how do we handle this in a way that is conducive to building accountability and commitment?

As leaders, we could beat up our teammates each time they fail to self-manage, execute a plan, or complete a deliverable on time. That's generally the first thing that comes to mind when we are determined to hold our teammates accountable, isn't it? But as you learned in the previous chapter, punishment is only effective for stopping unwanted behavior—it's not effective for promoting what we want. For that, we need to extend another invitation; we need to invite our teammates to *want* to act in a committed and accountable fashion. Again, this means that there has to be something in it for them to do so.

So what is in it for them? What's the payoff for putting in the extra work and living up to one's commitments and responsibilities? Quite a lot, really. When teammates keep their promises and honor deadlines, the level of frustration they have with each other diminishes, and productivity rises. Since everyone on the team is getting what they need from each other, they in turn are then able to perform at optimum levels, and when people are able to work at optimum levels, their chances of achieving individual as well as team success increases. As a result, trust among team members grows. As it does, they will increasingly view one another as a source of success rather than as potential threats. And as described in preceding chapters, there is a great deal of satisfaction to be derived from being able to focus squarely on results, rather than on how "certain people" are letting the team down. This is when people have those "ah-ha" moments—the realization that they can accomplish far more together than standing alone as isolated individuals.

So, what can a leader do to influence all of this? To be honest, it is not as complicated as you might imagine. As you institute the basics and begin to pull your team toward a delineated goal, start asking those "What do you think?" questions along the way. Even if the team displays reticence, urge them to put forth their ideas. Why is this so important? Most rational people know that they are not going to get their way all of the time, but they do want their ideas to be considered. As Ralph Waldo Emerson said, "Our deepest desire as human beings is to be understood." The simple truth is that when we invite people to feel heard, they are much more willing to buy in to the plan that emerges and commit to it, regardless of whether it was their idea or not. Why is this so? In part, because of the law of reciprocity. Simply put, this law states, that, "If you do something for me, in turn, I am obligated to do something for you." Whether we realize it or not, this law exists in all of our heads as an innate constant and cuts across all cultural and socioeconomic lines. At work, this law plays out in the following way: "Since you took the time to hear me out, I, in turn, will commit to the plan that is eventually adopted."

The other reason that feeling heard is so important is that people on construction teams *want* to be successful. And since they are anything but dead from the neck up, most construction people are usually more than capable of sifting out the best ideas from among the many on their own. But in order to do so, the ideas have to get on the table in the first place, and this is where a leader can exert his or her greatest influence. By creating a forum for ideas and making it okay to agree or disagree with what is being discussed, leaders create a culture where commitment can emerge naturally.

Conversely, if we choose not to extend this invitation, and instead ram our ideas down our teammates' throats, through their resulting silence, people will let it be known that they believe there are better, unexpressed ways of executing the job, and they will withhold their commitment until these ideas have been allowed to come to the fore. Reciprocity is also in play here, but in reverse; that is, "Since you didn't ask for my ideas or input, I don't have to fully commit to the plan that you have put forward." The added difficulty is that people won't be inclined to remain silent about their lack of buy-in for long—at least not to each other. At the coffee pot, or at the bar after work, they will seek out like-minded others who happen to agree with their particular divergent point of view. Rather than hitching their wagon to a shared sense of purpose, they form cliques and factions and, at this juncture, a project can unknowingly head off in different directions, creating multiple entrance points for waste to creep into the system.

Again, this is easy enough to prevent. Going back to our discussion about healthy conflict, by sending a clear message that the expression of dissenting views is not just okay, but vital to team process, this, paradoxically, aids in the formation of a unified team vision. When people realize that they can engage in healthy debate at the team level, the need to seek private alternatives diminishes. To solidify this sense, once all of the ideas have been fully vetted, and a course of action has emerged, ask each person to summarize the decision that was arrived at. This is a quick and easy way to make sure that everyone has drawn the same conclusion and will allow you to do some fine-tuning if you should find that there is some disparity between viewpoints.

Now, here is where accountability comes into play. Let's say that everyone was able to come to a decision on a proper course of action, and each person developed a work plan to satisfy the fulfillment of this overall plan. What happens if someone subsequently fails to deliver the goods? What do we do then?

We've already seen that ridicule or punishment is not the answer. On the other hand, we absolutely must call attention to the problems that failing to live up to a commitment can cause. But *how* we call attention to it is vital. It is the difference between coalescing a group of individuals into a team or inadvertently dividing them into factions. If you merely chastise a teammate for failing, you will miss the opportunity to invite people to want to be accountable and remain committed. In fact, you'll unintentionally train people to duck away from future responsibility. As we all know from our own bitter experiences, commitment and accountability don't happen in the absence of trust. So, if we are able to honor vulnerability and use failure as an opportunity to highlight the team's interdependence by effectively analyzing it, the whole team will see the value of acting in a committed and accountable way, and as a result, trust will actually grow.

Toyota utilizes a process that they refer to as the five why's to build commitment and accountability into their teams (Table 13.1). When a quality issue is detected or a deadline drops, the manager assembles the team and asks them *why* five times. This method is employed to identify problems upstream in the process and to teach the value of collective problem solving by identifying deeper countermeasures. Here is how it works: Suppose, in one of their plants, there was a puddle of oil on the floor. If we have no curiosity about why there is a puddle of oil on the floor, then all we will do is simply grab a rag, clean it up, and go on with our day. This is the most expedient countermeasure, but it does not get to the root of the problem. But what if we ask *why* there is a puddle of oil on the floor? This then leads

TABLE 13.1

Toyota's Five Why's Analysis

Level of Problem	Corresponding Level of Countermeasure
There is a puddle of oil on the floor	Clean up the oil
Why? Because the machine is leaking	Fix the machine
Why? Because the gasket has deteriorated	Replace the gasket
Why? Because we bought gaskets of inferior material	Change gasket specifications
Why? Because we got a good deal (price)	Change purchasing policies
Why? Because the purchasing agents are rewarded on short-term cost savings	Change the reinforcement policy for purchasing agents

us to a deeper countermeasure; that is, there is oil on the floor because a machine is leaking, so I'll tighten the machine. But this begs the question, *why* is the machine leaking? This, in turn, leads us to a faulty gasket that we can then replace. But if we ask *why* the gasket is faulty, this leads us to understand that gaskets of an inferior quality were purchased. And if we dig a little deeper and ask *why* inferior gaskets were purchased, we find out that it was because these gaskets were cheaper, and that the purchasing agents were evaluated (and received bonuses) on short-term cost savings. Do you see the point of this five why's analysis? Instead of attending to an array of apparent causes that just perpetuate the problem, by determining the root cause we are able to fix the problem at its source—thus preventing it from happening in the future. In this example, cleaning up the oil with a rag would have done nothing to solve the problem. Unless it is fixed at its source by changing the reinforcement contingencies of the purchasing agent (i.e., reinforcing purchases that lead to quality versus short-term cost savings) the problem will keep recurring.

We can analyze almost any problem in this manner. Let's go back to our mythical restaurant. The problem: it took over an hour for diners to receive their main course.

1. Why did it take over an hour to serve entrees to the diners?
   A: Because Andre needed to leave his station and help Jose with appetizers.
2. Why did Andre need to help Jose with appetizers?
   A: Because Jose had to leave his station.

3. Why did Jose have to leave his station?

   A: Because he ran out of scallops and had to go to the refrigerator in the basement to get more.

4. Why did this cause such a lengthy delay?

   A: Because when he pulled the scallops from the refrigerator, they were still frozen.

5. Why was this a problem?

   A: Because in their frozen state, the scallops could not be properly prepared, so the customers at the tables who ordered them had to be reapproached for an alternative—thus causing a delay that reverberated throughout the entire dinner service.

In this scenario, it would be easy to assume that Jose is the problem, but he is only a part of the problem. By asking *why* five times, we are able to discern the true root causes of the problem. Via this analysis, *why* Jose was struggling at his station is the truly important question. As it turns out, the head chef was also accountable because she clearly underestimated the amount of scallops needed for appetizers (planning error), which led to the execution error. Also, Jose should have let it be known that he was running low earlier (communication and judgment error) so that either more scallops could have been brought up and defrosted in time, or a decision could have been made to eliminate scallops as an appetizer altogether—before all of this had such a deleterious effect on the overall dinner service.

It is absolutely essential that the construction industry fully adopt the five why's methodology. Why? Because construction professionals spend far too much of their valuable time enabling or compensating for broken nonvalue-added procedural systems. As a result, they spend countless hours reworking flawed plans that inundate the entire process with waste. For example, a logistics group had the following agreed upon protocol in place: no PM was to present to an owner an approval letter (known as an A letter) for signoff if it had not been properly reviewed by a planner beforehand. The reason for this was clear: in the past, when proper planning was omitted, this usually resulted in costly reworks and lots of wasted man-hours. But in the name of expediency, a procurer submitted an A letter to the PM that had not been properly vetted. The good news is that in this particular instance, the courageous PM refused to pass it on to the owner. Unfortunately, rather than being applauded for her actions, the

PM was viewed as an obstructionist. But if her managers really wanted to have an efficient operation, engaging in the five why's would have netted them much valuable information. Here are the real reasons why this situation occurred:

The issue: The PM refused to pass on a letter to the owner.

1. WHY? Because it was not reviewed by the planner.
2. Why wasn't it reviewed by the planner? Because the planner's time was spread too thin by serving on other projects.
3. Why was the planner spread too thin? Because his boss (who was supposed to serve as a workload gatekeeper for him) was pulled away to chase new work and wasn't there to function in this capacity.
4. Why was an A letter submitted to the PM anyway, despite this being against protocols? Because the procurer felt pressured to produce it.
5. Why did the procurer feel pressured to produce it? Because the operations manager is required to generate a report of earnings projections to the corporate office based on A letters, and this report was now due!

Can you spot the absurdity of the root cause for this failure? The project could have incurred a large number of wasteful execution errors simply because everyone in the system felt pressured to satisfy a reporting responsibility required by their corporate office—something that had nothing to do with the actual building process. A nonvalue-added internal report was about to introduce waste into the process simply because someone felt pressured to break protocol. Thank goodness the PM had the courage to just say no!

This same method can be used for any construction issue. For example, "Why weren't we ready for inspection on the fourth floor?" "Why weren't we ready to pour on the east wing like we had planned?" "Why hasn't the submittal log been updated?" All of these issues can be probed via the five why's. I strongly encourage you to do so as a way of preventing costly quality errors that can repetitively occur on most projects. The key is to conduct this probe in such a way as to identify root causes rather than use it as a tool to pinpoint blame or identify scapegoats.

Another method you can use is something that I call *behavioral chain retracing*. This is a fancy term for working backwards from a failure point. It's similar to reverse scheduling. During a meeting, put a problem that has occurred on the far right-hand side of a white board, and ask the team to retrace why they think the problem occurred. For instance, you could write down "Missed pour date." Then people could start objectively filling in the reasons why this happened and in what order. For instance: "Didn't receive a timely response on RFI 621," "Didn't fully highlight the issue during a 'hot list' meeting," "Didn't send the request for information to the A&E in a timely enough fashion." Arguments will be made and positions jockeyed, but what will become abundantly transparent to everyone is that there were numerous failure points all along the way. Errors such as missed pour dates rarely boil down to just one weak link in the chain, as is readily assumed. They are usually the result of a cluster of errors. And like the KLM example cited previously, it will become apparent to all that there were multiple opportunities when any number of people could have stepped up, pulled the Andon lever, and averted disaster—but didn't.

The key to all of this is that this has to be done with the proper mindset, that all problems are team problems versus the sole domain of any particular individual—and that it is everyone's responsibility to fix them. Once a problem is identified, it's the team's job to analyze it, come up with a solution, and take responsibility to make sure that it doesn't happen again. If done well, people will take pride in participating in this process. Again, the challenge is for leaders to disarm their autonomic responses, because, believe me, when you are in this type of situation, physiologically, you will *want* to rake someone over the coals rather than engage in any problem-solving techniques. But if we let our emotions get the better of us during times of trouble, we will actually create a much bigger problem for ourselves and our teams to contend with—fear. After witnessing someone getting raked over the coals, the "innocents" may feel relieved that it wasn't them this time around, but in the back of their minds, they will be thinking about how to avoid shouldering any additional responsibilities for fear that, should a future failure occur, they would suffer a similar fate.

This next point may seem paradoxical, but it isn't, and I'm a stickler about it: deadlines need to acquire the feel of something sacred. Everyone needs to care about deadlines deeply and feel absolutely awful should one drop—even if they weren't directly involved in the issue at hand. If you have someone on your team who is complacent or blasé when deadlines are missed, dismiss them from the team ASAP. Notice the distinction here. When people try their hardest but make a mistake, don't chastise them for it. Instead, analyze the problem, figure out how to solve it and prevent it from happening in the future, and move on. But if someone, after being confronted, simply doesn't care, or dismisses it as unimportant, that is a horse of a different color. Then *they* are the problem, because, emotionally, they have already distanced themselves from taking any responsibility for the failure. In short, they aren't to be counted on in the problem-solving or continuous improvement process because they don't view their own attitude or behavior as part of the problem—so what is the point of keeping them on the team?

Similarly, it is incumbent on everyone, should they feel that they are in immanent danger of dropping a deadline, to let their teammates know as soon as possible. To promote this, it is vitally important for you to communicate this expectation *before* the project starts up in earnest. We are all human, and we will all fail from time to time. But by caring enough to let teammates know of an impending failure (versus covering their butts or pretending the problem isn't happening), and having a recovery plan at the ready, even in the face of defeat, each teammate can still make a significant contribution to the overall team effort. Such actions of accountability are symbolic invitations that teammates can extend to each other to renew their investment of trust in one another. While no leader wants to encourage mistakes or failures, it is vital to express your appreciation when someone has the courage to raise his or her hand to let others know that he or she is in danger of letting the team down. By rewarding such acts, you are not rewarding failure or complacency. To the contrary, you are reinforcing accountability since the person is attempting to head off bigger systems failures down the road by alerting others about potential problems up front.

Finally, there is one last leadership action you can take to increase accountability. Whenever possible, keep score. Whether you are tracking progress on the schedule, or a set of deliverables toward the fulfillment of a milestone or a goal, provide a form of visual measurement whereby the team can track its progress toward attaining the overall objective. It can be as simple as having a schematic color-coded chart at the building. Each time on activity is completed, progress can be tracked by changing the color in that area of the building to denote completion. Sound unnecessary? Think of how pointless the act of rolling a round ball down a wooden lane and knocking over a set of pins would be if we didn't keep score. Scorekeeping gives us a sense of accomplishment and meaning, and compels us to keep pulling, as a team, toward success.

14

Lean and Safe

This chapter is not intended as a guide for establishing a safety program at your site, as there are a plethora of great books on the market to help you do that. This section serves to highlight the leadership mindset required to make a safety program successful, and the pitfalls that can derail such efforts.

Safety programs are prevention programs. And like all such programs, they can be notoriously difficult to maintain, and success hard to measure. The reason is simple: we're asking people to do something that inhibits or inconveniences them in order to prevent something from happening that may never occur. A worker could potentially go on a site wearing nothing more than a tool belt and sustain nary a scratch throughout his or her entire career. Yet we require him to wear eye, head, ear, and foot protection as well as various and sundry restraint systems that slow him down or impede his movements all for the sake of preventing that one instance in ten thousand that could actually result in injury or death. In a very real sense, safety programs are an act of faith, because only on rare instances can anyone actually see the accident that was successfully prevented.

If you are reading these pages, you are probably already the type of person who feels a strong moral obligation to send people home with all their body parts working correctly. To meaningfully implement any safety program, one simply must care. But there is also a business side to safety; plain and simple, accidents have a monetary cost. There are the obvious costs of medical bills, lost time, and potential lawsuits. And there are also hidden or unrecognized costs, such as elevated insurance and healthcare premiums, and lost sales opportunities connected to inflated experience modification rate (EMR) rates.

In Lean terms, nothing kills productivity faster than a site closure due to a major accident, and this is does not just include the period between the incident and when the inspector decides to reopen the site. Weeks after the incident, the residual effects on morale and productivity linger as workers attempt to cope with the emotional ramifications of the accident. After all, that was someone's friend, poker partner, or best man who took that fall.

The question again becomes: How do we influence others to *want* to act in a safe manner? The obvious advantage of this is that if people *want* to act safely, then we won't have to be forced into the role of playing traffic cop all the time. Unfortunately, most safety programs are designed to be regulated by negative reinforcement (i.e., relying on the threat of punishment if workers do not comply with safety protocols). But as you'll recall, negative reinforcement paradigms have serious limitations. Most notably is the fact that if the enforcer isn't around to deliver the punishment, people will revert back to risky behaviors in the same way that drivers speed up on the freeway as soon as they see the police car pull off at the next exit.

So how can we influence positive change and get people to want to do something that they often don't see the value in doing? There is a great story about a dentist who was a master at this. A patient with gum disease was complaining to him about how inconvenient it was to maintain a flossing regimen, in an attempt to manipulate him into saying that it would be acceptable for him to take shortcuts in his dental hygiene. The dentist listened, thought for a moment, and then said, "I have a solution that I think will work for you; just floss the teeth that you want to keep."

Unfortunately, by law, we can't present such options to our workers. But I did hear a general foreman use a similar line of reasoning with excellent results. After witnessing and ordering down an unsecured worker who was reaching out precariously to tighten a bolt while standing on a beam fifty feet off the ground, he said, "Let me ask you something. Are you nuts, depressed, or do you have so little respect for your company, your family, and yourself that you don't care if you get hurt?" After several mumbled excuses, the general foreman continued: "So here's your choice; I either give you the number for a shrink to get your head on straight, or you never fail to wear your restraint harness ever again! What's your choice?" What I liked about this approach was that the general foreman didn't rely solely on delivering a punishment to get his message across. He was also trying to penetrate the worker's thinking by getting him to understand the implications of his actions and take responsibility for making unsafe choices.

Whenever possible, it is important to engage in frank open dialogue, akin to *the five why's,* whenever noncompliance is observed—particularly when these behaviors deviate from how people had been acting. Let me refer back to my days in the corrugated box industry to illustrate this point. Seemingly out of the blue, a plant was experiencing noncompliance issues with eyewear protection protocols. The reason they called me was that their policy was very straightforward; each time a worker was caught not wearing eye protection on the floor, he or she was written up, and three write-ups resulted in termination. Per their union contract there were to be no exceptions. Unfortunately, at this point of their enforcement, they had already handed out two write-ups to some of their best employees. They were clearly at an uncomfortable and precarious crossroads. But as suggested in *The Toyota Way*, in instances such as these, it is much better to stop the line (*Andon*) and analyze why the problem is occurring, rather than continuing to charge forward. After a cursory assessment, it became clear that in their zeal for compliance, the management team failed to ask this one simple question, "Why aren't you guys wearing your glasses now?" When they finally asked the question, here is what they heard:

- "They pinch my nose! You try wearing these damn things for eight hours!"
- "They fog up! It's more dangerous for me to wear them around the corrugator than not to!"
- "People make fun of me! I look stupid in these damn glasses!"

It didn't take long for the management team to get to the heart of the root cause for noncompliance. It wasn't that their workers had suddenly become jerks. In an effort to cut costs, the plant manager had decided to purchase a cheaper brand of glasses. But the actual cost of this choice was now apparent. Though he was able to save 46¢ per pair, when the costs of potential OSHA fines, possible injuries (while wearing the glasses), and the potential loss of key employees were factored in, the "fix" for this problem was extremely simple—junk the new glasses and bring back the old, slightly more expensive ones that performed much better.

Adding to this point, T. J. Lyons, a regional safety manager in the northeast for Turner Construction, stresses the importance of not underestimating factors that may appear unimportant on the surface. He was once asked to lead an extrication team for a local fire company, and bought small bump caps, Kevlar gloves, and goggles—exactly what was needed. The men would not wear any of it. But when he purchased NASCAR style gloves, wraparound glasses, and rescue helmets like Gage and Desoto wore, he couldn't keep the gear off the guys. As T. J. often says, "Safety isn't about winning the battle of wills—it is about getting people to *want* to act in ways that are safe."

In fact, the use of punishment alone is only appropriate in situations when dangerous actions are observed and potential harm is imminent (the unsafe behavior needs to be stopped immediately). But this should always be followed by a coaching session, unless a worker has been previously warned. In such cases, termination or expulsion from the site may be warranted.

While it is natural to get caught up in being an enforcer, don't neglect one of the most important tools in your arsenal—praise and recognition. To transform a culture, punishment and negative reinforcement only go so far; people need to experience a benefit for engaging in safe behaviors. Try using these simple behavioral tools:

- Catch people doing something right (i.e., when they are in compliance) and call them out on it in a positive way. (If we don't, we run the risk of putting safe behaviors on an extinction curve toward elimination.)
- Ask subcontractors to recognize the safe behaviors engaged in by other subcontractors, that is, keeping a work area especially clean. (It feels good to be recognized by your peers.)
- Celebrate team successes for sustained periods of compliance. (Nothing breeds continual success like collective recognition and reinforcement.)

From a Lean perspective, try seeking out safety practices that can actually *increase* efficiencies and productivity while at the same time providing workers with proper safety protection. Unfortunately, in so doing, you may run into some hardheaded individuals who may not agree with your line of reasoning. Here is an example:

A safety manager watched a plumber climb a ladder four times to solder on a connector. He had eighty to go. He could have worked faster from a scissor lift, and he would have been comfortably restrained within a passive system. When the safety manager asked him why he wasn't using the lift, the worker simply replied, "Ask the boss."

At times such as these, a little diplomatic cost-benefit reasoning with higher-ups can go a long way.

This leads us to back to the most important tool in any leader's safety arsenal: his or her own attitude and resulting behavior. As a leader, there are opportunities that you have each day to increase the chances that the people on your job will act safely and adopt a safety mindset. Here is a simple checklist to help you on your way:

Yes No Do you lead off every meeting with a sincere review of safety issues, noting not just noncompliance, but especially strong safety compliance as well?

Yes No When you go out to the field, do *you* wear all the required safety apparel, thus modeling the proper behavior for everyone else?

Yes No Do you reward people for *not* making an exception for VIPs and visiting dignitaries? (There was a great story in San Francisco of a safety person who refused to let Mayor Gavin Newsome go on site without a hard hat because he complained that it would hurt his hairdo. The safety person was verbally rewarded for his ethics.)

Yes No Do you follow the five progressive rules of safety: (1) eliminate the hazard, (2) substitute with something less hazardous, (3) isolate the hazard from the worker, (4) reduce the worker's exposure time, and (5) and provide personal protective equipment?

Also, make sure that your safety practices are grounded in empirical data. There is a now famous story of a group of construction workers who were seen walking around Manhattan wearing life vests. They were doing so as protection against falling into slurry walls that were under construction. It was their belief that the vests would save them from drowning should they happen to fall in. When the lead person was asked by a safety manager how he knew the vests would work, he shrugged his shoulders and said, "Someone told me." Subsequent empirical testing quickly revealed that this was a faulty belief. As the vests became saturated, they actually pulled test subjects deeper into the slurry. As it turns out, the only true preventative action to keep from drowning in slurry is to prevent the fall from happening in the first place. So here is an important lesson to take forward from this example: the last thing we ever want to do is to inadvertently encourage risky behavior by giving workers a false sense of security by encouraging the use of a flawed device. Not to mention that this also damages our credibility when trying to institute practices down the road that actually do work.

One final point: Never underestimate the power of a well-crafted letter to not only reinforce the safety behaviors you are looking for, but also acknowledge efforts toward continuous improvement.

Turner Safely Building the Future

January 5, 2009

Grau Contracting
Jerry Sheridiane
3300 Panel Way
Saint Charles, MO 63301

RE: Paul Reeves, Foreman—Xanadu Project

Mr. Sheridiane,

Early last year I had the opportunity to meet Paul Reeves, your project foreman, at our Xanadu site in New Jersey. He was overseeing several crews lifting panels. We spent considerable time reviewing rigging, how it should be inspected and maintained. Paul provided some insight that became an opportunity to Turner.

That simple conversation started an effort across the Northeast to take a look at rigging in an effort to reduce or eliminate risk from our lifting operations. In 2008 we inspected over 41,000 pieces of gear. The results were a new, lasting focus at Turner on the value of great rigging to eliminate this "weak link" from our hoisting operations.

I just wanted to recognize Paul and the effect of a simple conversation with a professional rigger.

Sincerely,

T. J. Lyons, CSP
Regional Safety Director—NE
Turner Construction

And finally, here is a letter T. J. sent to a worker's home. There is a simple beauty in this; not only does the worker receive a positive reinforcement from his place of work, but he also is also likely to receive one at home as well. Or as T. J. puts it: "This is one of the best motivators I have found. His wife will wonder all day what this letter is about, and when he finally opens it, she will see what he does and what he cares about both at work and in his life."

Turner Safely Building the Future

September 10, 2006

Rick _____—Project Superintendent
1234 ABC Way
Anywhere, USA

Rick,

In a conversation last week with my site safety coordinator, Andrew Leone, he noted that your team is doing a fantastic job on the Mills Project. I also agree, we watched you guys last week and you do the routine safety efforts—routinely. Having worked with you guys in Reno, I was not surprised.

In fact, Andrew noted: "Exceptional efforts in conducting proper fall protection, flagging off hazardous work zones, and creating an atmosphere where practicing safe productivity is part of the business strategy. Through his efforts, Rick has proven that safe productivity can be accomplished in conjunction with maintaining a rigid work schedule."

For years I have been pushing the need to show that safety and productivity are related. You are one of the few that has figured it out.

I speak for the entire Turner team in saying thanks.

15

Fine-Tuning the Line: Keeping Your Fingers on the Pulse via Continuous Assessment

> The most important and lasting lessons that I learned from Bigfoot were about personnel and personnel management—that I have to know everything, that I should never be surprised. He taught me the value of a good, solid, independently reporting intelligence network, providing regular and confirmable reports that can be verified and cross-checked with other sources. I need to know, you see. Not just what's happening in my kitchen but across the street as well…
>
> **—Anthony Bourdain,** *Kitchen Confidential,* 101

Your best efforts at building a high-functioning team will go for naught if you don't keep your fingers on your team's pulse on a regular basis. The reason? Any hiccups on the people side that require midcourse corrections that go unnoticed can quickly lead to waste on the technical side. The one thing you can count on for any construction team is that its dynamics will always be in a state of flux. So, as Bourdain suggests, you need to remain aware and assess any potentially detrimental changes in team chemistry.

What I mean by the term *assessment* is the willingness to see things as they are—dispassionately and objectively—and without distortion (i.e., interpreting negative information as a personal attack or, conversely, believing it to be better than it is). This is a lot easier said than done. To survive life's slings, arrows, and recriminations, we all live, to some degree, in a world of self-deception. We simply couldn't function if we kept the harsh realities of all of our shortcomings in clear view at all times. So, to a large extent, self-deception is natural and, within reason, even healthy.

But managers who are able to demobilize this natural defense mechanism and get comfortable looking reality square in the eyes stand a much better chance of helping their team refocus should they temporarelly lose their way.

Assessment, when conducted properly, is the opposite of ignoring issues or simply emotionally reacting to problems as they arise. It is the means by which we get our facts straight so that we can give a set of revised and accurate sailing orders that will help our team reach its desired destination.

There are three primary sources of assessment data (indicators) that you need to gather on a continuous basis: *behavioral*, *procedural*, and *external*. Any of these can serve as red flags or precursors to potential productivity breakdowns. It is important to note that at the point of detection, the cause that lies beneath the red flag will likely be unknown. This is the whole point of continuous assessment: to be able to ascertain the warning signs of a root cause *before* the issue becomes a costly problem.

BEHAVIORAL

Behavioral indicators are things that you can directly observe—in other words, what you can hear, see, and feel. They include voice volume and inflection, body language, and facial expressions. This is information that people communicate indirectly, usually because they are uncomfortable conveying it by more direct means. The reasons people choose to communicate indirectly are multifaceted. Some do so because they are preoccupied with how they will be perceived by their boss and don't want to be labeled as either whiners or complainers. Others may be afraid of how you will react to bad news (kill the messenger syndrome). And still others will simply assume that since you are the boss, you already know about the problem and are just choosing not to do anything about it. The latter is one of the biggest team fallacies that exist in the workplace. In reality, most data that reach a manager's ear are highly filtered, that is, they hear what others think they *want* to hear, and this filtering worsens the higher up you are in the organization. By paying attention to behavioral cues and digging deeper to discover their root causes, you will be able to identify what is frustrating your team so they can get back to focusing on results. Pay particular attention to the following:

Changes in behavior. When an even-keeled person starts lashing out or looks like he is carrying the weight of the world on his shoulders, it is time to pull him into your office for a chat. Not a scold, a chat. Remember: the goal is to find out what's going on and why he is in such discomfort. When you do this, some people may initially become defensive. But soon he will be touched by that fact that you took the time to notice changes in his mood and will be genuinely grateful (and relieved) for the opportunity to get what is bothering him off his chest—before having a complete meltdown.

Sighs, eye rolling, head shaking. These behaviors are usually most prevalent during staff meetings. If you see them, it is time for you to stop your spiel, put down your note pad, and say, "Okay, folks—I'm looking around the room and I'm seeing a lot of things that tell me that people aren't pleased right now. Talk to me. What's going on?" For example, I once observed a group of engineers snickering and making sarcastic remarks among themselves whenever anyone from accounting or the field spoke. They had no idea just how obnoxious and self-serving this appeared to the rest of the team until they were confronted. With a little probing, it turned out that contrary to their appearance, they were actually very concerned about the project. But they felt that their considerable worries about the buy-out were being blown off while the team overfocused on more trivial issues. The manager did a great job of validating their concerns, while at the same time pointing out that the whole team would be better served if the engineers voiced their worries directly, rather than through sarcasm or side talking.

Finger-pointing. When teammates begin to cannibalize one another (blame each other for team failures), this should get your attention immediately. Allowing this to go unchecked or to fester will send a poor message to the rest of the team and will have disastrous effects on productivity. Your constant mantra should be: "We can disagree; we can even be passionate about it, but we all have to be willing to do whatever it takes to get the issues on the table and work toward a solution." Often, issues that escalate to finger-pointing boil down to fairly manageable problems provided that they are caught early on. Such things as poor role delineation (i.e., "I thought you were doing that!" "No, that's your job!") can easily be corrected by simply slowing down the process and clarifying who is to do what by when.

At other times, the root cause of the finger-pointing may indeed be due to the fact that someone did not live up to a promised deliverable. But as in the case of Jose at the appetizer station, there is usually a communication component that also broke down and led to a breach of trust (i.e., the person who let the team down knew he was about to fail, but chose not to communicate it to the rest of the team). So, when the full impact of the dropped deliverable finally hits home, those most affected reacted harshly and began to question the motivations of the teammate who let this happen. This type of situation will require you to sit people down, lend them your calmness, and work the issue through to resolution. For example:

Look folks, we're all human. The fact is, even though none of us want to, we're going to let each other down from time to time. But we have to work on ways to be smarter about it. Tommy, in the future, if you think you are going to blow a deadline, you need to go to Sally beforehand, and let her know where you are at and what your recovery plan is so she can make the necessary adjustments to her own work plan. Understood? And Sally, in the future, rather than blast Tommy, give him the opportunity to come to you with a recovery plan. Remember, everybody, we are here to help each other. Right now, we're working against each other. Let's bring these kinds of things up early so they don't bite us in the butt later.

Palpable tension. I can usually tell the minute I open the trailer door how things are going on a team simply by how I am greeted. When the tension is so thick that you can cut it with a knife, it's time to stop and have some discussions about what you are seeing. Again, the goal is to find out what is frustrating people and make it okay for people to talk about it. But first, you'll have to assure people that what is said won't be used against them in the future. Usually this is best done individually, rather than in a group. But it should be announced to the entire team beforehand: "Look folks, I can tell everyone is tense as all get out. I can see it in your faces. So, I want to spend some time with each of you to find out, from your perspective, what's going on so we can move forward."

Avoidance. People avoid taking on new tasks and responsibilities like the plague. No one wants to do anything that could put him or her in the line of fire. This is an indicator that trust on the team (willingness to put oneself in a vulnerable position) is fairly low.

Clock watching, early departures, late arrivals. People in construction want to be successful, and they will often put in long hours to be so. But if they feel thwarted in their efforts, they'll start voting with their feet. At first, they'll put in their time with little passion and then go home. But then you'll start seeing people trickle in late or leave early. This is the point at which those who had no intention of returning a head hunter's phone call will begin entertaining them.

Increased needling, wisecracking, and sarcasm. I'm not a big believer in political correctness in either my work or private life. I enjoy people who are quick-witted and funny. But I also believe that people should treat each other with respect. Keep an ear out for when humor becomes a thin veneer for underlying anger or frustration. Too much sarcasm shuts people down and destroys vulnerability.

Increased use of email. When people within close proximity begin to rely on emails to communicate their displeasure with one another, it is usually an indicator that unhealthy conflict has crept into the team dynamic. It is important to interupt such exchanges early on so they do not escalate.

The only sound you hear in the trailer is the clicking of computer keys. Job sites should be dynamic places. People should be asking questions of one another, pouring over drawings, and hashing out iocues via open dialogue. The sound of silence, in short, should scare the pants off of you.

Any of the above are indicators that it is time to take your team's temperature in a serious way. Again, don't be afraid of what you will find out. Be glad that the issues are coming to the surface now rather than farther down the road, when they will be much more costly to fix. Just draw your team out by sharing what you're observing, show concern, listen, and see what emerges. The solutions are usually pretty apparent once you've taken the time to fully listen and understand the issues. The only ways that you can blow it is to become defensive, dismissive, controlling, or verbally combative.

LISTENING SKILLS

I'm not going to launch into a long discourse about effective listening. There are already many excellent resources to help you with this. (I particularly like *Listening: The Forgotten Skill: A Self-Teaching Guide*, by Madelyn Burley-Allen.) But I will say this: if you can learn to master this one simple skill, there will be few people issues that arise that you can't put right. All it requires is a willingness to slow down (*Andon*) and to understand the issues and concerns from the *other person's perspective*. This doesn't mean that you have to necessarily agree with the other person's point of view. You just have to understand it.

For instance, let's say that you've noticed a change in one of your employee's behaviors. (They've gone from being as giddy as a schoolgirl to looking like they have the weight of the world on their shoulders.) Step 1 is to carve out some uninterrupted time where you can meet with the person privately. (Remember to turn off your phone, get away from your computer screen, and close the door.) Then, start out by sharing your observations. Don't make it an interrogation or try to control their responses, just let them know that you are there to help. "I've noticed that you are not your usual self; is there anything going on that I can help with?" If the other person welcomes your inquiry, try gathering some facts by asking probative questions, such as

1. What's going on as you understand it?
2. What is happening that shouldn't be?
3. What isn't happening that should be?
4. How long has this been going on?
5. That sounds frustrating. What have you tried to do to make it work?
6. Have you had any success in getting it to work?
7. If things were going perfectly, what would it look like?

These questions are designed to help the other person slow down and view the situation more dispassionately. Once you've gathered the facts, and both of you feel like you've adequately defined the situation, take the next step to help move things forward. These are what are referred to as action questions, and they sound like this:

1. In your opinion, what specifically would help to improve the situation?
2. What would be the first step you would take?

3. What help do you need from me? From your teammates?
4. What's at stake for you if we don't get this issue to move?
5. What resources do you have/need?
6. Is there anything that you feel you could be doing to inadvertently contribute to the problem?
7. Are we in agreement in terms of what a positive outcome would look like?
8. Whom else might we need to pull into this process?
9. When will you start?
10. Let's agree on a time to do some follow-up to see if your plan worked.

Action questions are designed to help the other person formulate a plan to address his or her own concerns, rather than you having to take it over for him or her (unless they agree that it would be appropriate for you to do so).

You'll be amazed at what little effort is required to help move a seemingly impossible situation forward. All it takes is a little time, patience, and skillful guidance.

Some managers actually believe that acknowledging team problems is tantamount to admitting weakness. But the opposite is true. When you are open to feedback and able to fully listen, you are announcing to your team that you are willing to do whatever it takes, including hearing difficult feedback about yourself, to help them improve. Acknowledging the existence of problems actually creates hope. It indicates to your subordinates that they won't have to fight to get heard, and that there is no agenda other than seeking an optimum project outcome. Remember: Effective leadership is not about being perfect—it is about successfully adapting to the challenges that arise.

PROCEDURAL

RFIs and submittals not properly vetted or logged, PCOs or change orders failing to meet quality standards, schedules not followed or updated, safety protocols not adhered to—all of these maladies, at first blush, may appear to be the handiwork of one or more nonperforming individuals. But more often than not, they are the result of a team process that has broken down. This is not to say that you can't have a bad apple or two in the mix. But procedural failures such as these often have broader root causes, such as people not being properly trained ("I don't know how to do what you're asking me to do"), people too overwhelmed with too many tasks ("I know that it's important, but so is everything else I'm doing"), poorly delineated roles and responsibilities ("I didn't know that that was mine to do"), or coordination issues ("I can't do what I need to do until I get _____ from _____!").

When properly assessed, the solutions to these problems are more accurate and sustainable, and won't require a Stalinesque purge to rectify. Again, this is the whole point of conducting a thorough assessment in the first place—so you can focus your energy on where it is needed (i.e., providing procedural training or prioritization skills training, conducting a brief ten-minute "What's hot for me today" meeting at the beginning of each day to spark coordination, etc.). These fixes will often prove far more useful in the long run than becoming obsessed with weeding out a procession of less than A+ players. After all, even if someone in particular was to blame for a certain failure, the issue still needs to be addressed, and the team needs to put their collective heads together to figure out the best ways to make that happen—and prevent similar problems from happening in the future.

There is one important caveat: if you clearly have a team cancer in your midst (i.e., someone who consistently dumps his responsibilities on others, fools around when he should be working, or repeatedly refuses to take responsibility for mistakes), don't put this on the team to solve. Deal with this person personally, swiftly, and harshly. He needs to get the message that high standards are expected of everyone. No exceptions. No excuses.

EXTERNAL FEEDBACK

External feedback is anything that you hear—directly or indirectly—from architects, owners, subcontractors, end-users or designated community contacts. Often, this feedback will be cloaked in avoidance or artificial niceness, so it is especially important that you pay attention to subtleties, such as

Breaks in chain of command. Any breaks in the expected chain of command should get your attention immediately. For instance, let's say a general foreman (GF), who should be interfacing with the general contractor's (GC) project superintendent (PS), instead, continually seeks out an engineer. There could be a number of reasons why this is happening, but it is important for you to specifically find out why. Perhaps the GF doesn't trust the PS, and is seeking out a more reasonable person to work with. Or, maybe the GF thinks the PS is incompetent and he's discovered that the information he receives from the engineer is far more reliable. Or, maybe the GF has discovered that the engineer and PS rarely speak and that he can exploit the engineer's inexperience to his company's advantage. Whatever the reason, you'll need to check it out:

"Hi, Terry, can I talk to you for a second? I've noticed that Jim from ABC Plumbing always goes to Laura in engineering rather than to you. Is there a reason for that?"

"Yeah, I got tired of Jim's constant whining about what he can't do. So I asked Laura to deal with him for a while."

If you get feedback like this when you check things out, you can breathe a sigh of relief. It means that your PS and PE are talking, strategizing, and have things well in hand. But if Terry was not aware that this was happening, or he was aware and has been secretly stewing about it, it's time to sit both parties down and work this through.

Scrutiny. Whenever the owner, the architect, or anyone else starts picking apart your schedule or seriously questions your means and methods, you can bet dollars to donuts that they have heard something through the grapevine that has tweaked their anxiety. Rather than talking around it or trying to control it, try this instead:

Excuse me, Marsha, but I can't help noticing that that you keep suggesting that we're not going to have the raceways in the east wing finished on time. Are you like me—just a natural-born worrier—or has something specific come up that has you particularly concerned?

By speaking directly, you signal your willingness to hear your external partners' worries and concerns, and assert that you aren't merely seeking to mollify, ignore, explain away, or brush off them off. This invites vulnerability and breads fruit with your external partners.

Increase in the number (or a change in tone) of emails. If, as a project leader, you begin to receive an inordinate amount of emails from external partners, this should get your attention. It could mean any number of important things. Perhaps, they do not have a strong grasp of the organizational structure and are therefore blanketing everyone on the project in the hopes that their concern will eventually find its way to the right person. Or, they may be attempting to signal their displeasure or lack of confidence for a particular individual within the structure. If this is happening, or the tone of emails becomes more CYA or accusatory than usual, it is important to respond immediately with a phone call or face-to-face meeting, as such emails are often the precursor to sending positioning letters. By carving out a time and place to engage in productive conflict, you can often head off much nastier, formalized notices down the road.

Please don't respond to a pointed email with an email of your own. For some reason, people say all sorts of nasty things in emails that they wouldn't dream of saying face to face or on the phone. A retaliatory email, while feeling justified, just ads fuel to the fire. It is better to break this cycle, picking up the phone instead, and find out the root cause of the email angst.

ASSESSMENT AND THE BASICS

In Chapter 6, I alluded to ways that you can utilize your organizational chart to diagnose team problems. Here is how this works. Let's say you've been taking the pulse of your team and you've gotten some rather pointed feedback:

"The chain of command is violated with alarming frequency."

"Whose project is this? The PX, PM, or PS? You all seem to be struggling for control of the project."

"Cost is a mystery. If we have controls in place I have no idea what they are."

"Twenty-five RFIs are over fifty days old—I found this out from one of our subs."

"I heard there are 187 unique contract requirements. What are they?"

"I know that I have to issue a PCO, but I have no idea what the quality standard for a PCO is."

"Changes are not incorporated into the contract documents. Who is in charge to make sure this happens."

"Procedures seem to always be in a state of evolution. When can I expect them to fully evolve so I will know what I'm doing?"

As the leader, it will be important for you to take a step back and objectively ask yourself what side of the house these problems are occurring on. In this example, if you said engineering, you are dead on. Now, you'd be tempted to jump to the conclusion that somebody on the engineering side wasn't doing his or her job, and this may indeed be the case. But often problems such as these are a symptom of a much bigger problem. Examine your organizational structure. Are there any missing players or weak links in the structure? Are there any unintended bottlenecks that are created by the structure? In the above example, a brief examination of the project organizational chart revealed that the team was attempting to function without a lead senior project engineer. As a result, the project manager (PM) was attempting to fulfill both roles and was predictably failing at both. Ironically, the management team was well aware that this key role was vacant. But they were not fully cognizant of the impact their continuous intrusions into each other's work, in order to plug procedural holes, and their lack of an overall execution strategy, was having on the team until they solicited their feedback. The reality was that this team stood no chance of success until the managers figured out a meaningful and systematic way to fill the voids created by the missing lead engineer (i.e., clearly dividing up the tasks in a well-publicized fashion, each manager formally stepping down into a designated role, elevating a junior engineer into a senior role, or lobbying the operation's manager to get them a lead engineer).

Let's take a look at another example—this time, at a company-wide level. An electrical contractor was experiencing problems that were preventing them from being the kind of company that they wanted to be: a flexible, nimble organization, able to quickly and readily adapt to rapidly changing market conditions. Their problems clustered around a recurring set of complaints:

- They did not share human resources well. (VPs often hoarded their best PMs and GFs rather than assigning them to where they were most needed.)
- Sales opportunities were lost rather than shared. (If one group lacked capacity, they hoarded it anyway. But since they lacked the capacity, they often failed to act on the opportunity in a timely fashion and failed to submit a bid in time, thus losing the sales opportunity for the entire company.)
- Some of their better managers clearly didn't have enough to do and were bored.
- Some divisions had too little oversight and their jobs became money bleeders.
- VPs often had to drop down to clean up problem jobs.

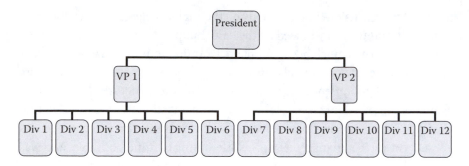

FIGURE 15.1
Restrictive structure.

Figure 15.1 shows how they were organized. Can you spot the problem with the organizational structure? This contractor was actually set up as if it were two competing companies. Paired with their bonus structure that rewarded individuals on specific projects, there was no incentive for anyone to share opportunities or human resources; in fact, there was a disincentive to do so. In addition, there were far too many divisions and projects for either of the vice presidents to properly oversee. If they had a manager in place who was incompetent or corrupt, they usually only found out about it when the job was already bleeding money.

During a strategic planning session these problems were posed and the managers were given the opportunity to create a new organizational structure to address the issues. Figure 15.2 shows what they came up with. This new structure not only allowed them to share resources and opportunities,

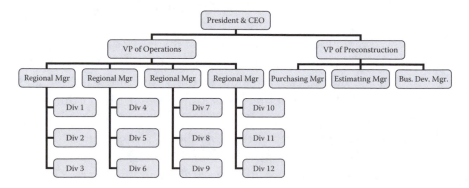

FIGURE 15.2
Flexible and nimble structure.

but also it compelled them to do so. With one VP overseeing all of operations, there was no incentive to hoard people. And with the other VP overseeing preconstruction, there was no disincentive to spread business opportunities to divisions that had the capacity to execute them. Paired with a new shared bonus structure that rewarded everyone for overall company performance, there was actually an incentive to share resources over the entire company. And by creating a new layer of empowered senior managers, the problem managers or projects under them quickly came to light and were addressed well before the company sustained negative monetary impacts.

I don't want to oversimplify the situation, or mislead you into thinking that all their problems were solved by simply modifying their company structure. They also overhauled their training program, developed a go/no-go strategy for venturing into new markets, and geographic regions developed alternative product lines, focused on getting the types of projects that they did well and were profitable doing, and redoubled their recruitment efforts. But what is certain is that they would never have grown from a $250 million company to the $900 million company they are now in a five-year span if their organizational structure had remained the same. And they would not have been able to change their structure if they hadn't been willing to honestly assess their problems.

ORGANIZATIONAL CHANGES AND THE ROLE OF EMOTIONS

If, based on your assessment, you find your team wanting or in need of change, don't underestimate the role emotions (particularly status anxiety) can play in the success or failure of the resulting organizational structure. By way of a scenario, here is how this can play out:

> After walking the job, and looking over the schedule and budget, the OM pulls you (the PM) aside and says, "I know that I didn't give you the strongest team in the world, but I think the way this team is organized isn't working." You actually agree. After seeing them in action for a few months, you have a lot better handle on everyone's strengths and weaknesses, and recognize that you have more than one person who is "playing out of position."

After grabbing some dinner, you and the project executive (PX) review the organizational chart and start making changes. You "lower" a superintendent who is new to the company from a lead role—because he is struggling mightily with the paperwork—and place him under a more senior superintendent, believing this to be the best way for him to learn. You shift a person who has a strong engineering background back to engineering from the field. You elevate an engineer who doesn't have very good people skills to a lead role, because you think, despite her deficits, the young people on the team will benefit from her technical expertise. You then decide to go one step further and shift around a number of other people's roles and responsibilities to better suit their strengths. It's 11:00 p.m. and you and the PX sit back and admire your handiwork. You are both pretty excited about the new structure and feel in your gut that it will work well.

The following day you roll out the changes to the staff. But as you do, you are stunned. Instead of excitement, people are staring at you, wearing rather stony expressions. You can tell that some people are actually fuming with anger. You can't believe this! Now you're the one who is getting steamed! "What an ungrateful bunch of selfish schmucks," you say to yourself. "Do they have any idea how much work the PX and I put into this? Look at them! They don't care about the project! All they care about is themselves! Maybe it's high time that I started shipping some of these bozos out!"

Wow! So, how do you account for your staff's reaction—and yours? Why do you think that something that you and the PX undertook with the best of intensions suddenly went so horribly wrong? And how was it that this emotional upheaval occurred in a mere matter of seconds—after hours of careful and rational planning? You guessed it: unwittingly, you activated the team's and your own fight-or-flight response.

You see, it doesn't matter that you had just laid out the most brilliant organizational scheme known to mankind. Nobody in the room heard a word after you said, "So, we're going to change things up a bit." All they heard was that the secure little world that they had known and loved—however dysfunctional it may have been—was changing, and they felt threatened by it. The rationale at that point didn't matter. What they were saying to themselves as they were glaring at you was

"Hey, I liked what I was doing, and I thought I was doing it well. Are you trying to tell me I'm doing a bad job?"
"I just finally figured out what I was supposed to do, and now you are changing my assignment without asking me? This is so unfair!"

"The OM promised me I'd get a chance to work in the field! Who are you to put me back in engineering!"

"No way! You just put me under somebody I hate and have no respect for! This is bullshit!"

"How come none of us saw this coming? How long have you been planning this? What other changes do you have in store? Are you going to start firing us next?"

And as you quickly found out, you are also not immune to the effects of status anxiety. Despite those letters after your name, you're a human being too. Admit it: you were expecting people to be as jazzed about these changes as you were—maybe even pat you on the back and laud your organizational brilliance. Instead, you got walloped and your own autonomic responses kicked in. In essence, you just worked your butt off on something with the best of intentions and got punished for it. And in that moment, within milliseconds, the very people you were trying to carefully craft into a solid team suddenly became a threat to your existence—and you responded in kind.

Here's the point that is absolutely essential to grasp. In these types of situations leaders (particularly engineers) rely too heavily on their technical analysis to carry the day and grossly underestimate the power of emotion. The fact is that people are and always will be emotional animals first, and rational, cognizant beings second. When making these kinds of changes you have to anticipate and accept the fact that a great deal of status anxiety will come with it.

And here is another vitally important reason to attend to the emotional side when addressing organizational changes, and this is particularly true for changes done on a broader scale (i.e., business unit or company-wide). During such transitions, you run the risk of losing your best people. Let's be completely honest. Your marginally productive people aren't going anywhere. They may grouse and complain, but you and they know that their options are limited. But your top performers always have options—even in down economies. You want to make sure that they are the first to know that they have a place within the new organizational structure.

In such situations, it is better to lay the groundwork for change by asking for feedback from all of the staff about what is and isn't working and asking their ideas for improvement. Done in this manner, the resulting changes will usually be in line with what you have already been thinking about and will feel reasonable—even logical—to everyone else.

But if you have fallen victim to what a colleague of mine calls *premature announciation,* here are some tips to help mitigate the damage. First, don't get blown away by people's initial adverse reactions. They are normal and natural. People are creatures of habit and change throws them for a loop. Anticipate some angst, roll with it, and don't take it personally. Second, build in some time (a week or two) for people to make the mental transition. During this period, make yourself available to answer questions and to allow the staff to express their worries and concerns. By the end of the first week, 85% of the staff will be ready and able to smoothly make the transition. Again, honest discourse goes a long way toward helping people get past their initial gut reactions. Also, stay open to receiving feedback. Your staff may have some very valid input that might make you rethink some of the changes that you have made.

For the other 15%, give them a little more time—another week or so. But if they continue to buck the changes or begin to act out their displeasure by engaging in undercutting behaviors (withholding information, gossiping, rumor mongering, excessive bad-mouthing of management, refusing to accept the new assignment, refusing to report to the person they have been assigned to), you will have to sit them down, point out the negative effects their behavior is having on the team, and ask them to make a choice. They are either going to have to find a way to live with the changes, or you are going to need to find another home for them. This isn't a threat; it's just the reality of the situation. A new house divided—no matter how much they may not like the new house—cannot stand.

PAYING ATTENTION TO THE GOOD STUFF

So far, we've limited our discussion of assessment to scanning for potential problems. But let's not forget about the other end of the spectrum. It's also important to assess what is going well and take the time to reinforce it with praise and recognition. Pay attention to such things as

- Decreases in the number of rejected or dismissed RFIs by the architect for insufficient vetting
- Decreases in the number of outstanding PCOs, billings, etc.

- Increased rate of timely billings
- Decrease in the time it takes to find current drawings
- Increase in the number of job walks taken by PMs or engineers with field personnel
- Increase in the number of overall interactions in the trailer (subsequent decrease in the amount of time people spend working solo at their computers)
- Actual compliments (verbal or written) from the owner, A&Es, or city officials
- Increase in the number of questions asked in the staff meeting
- Decrease in the number of failed deadlines
- Decrease in the percent of failed tasks on work plans per week
- Increase in staff to volume ratio
- Increase in overall productivity rates
- Decrease in the amount of times people seek you out to complain about others (with a proportional increase in time that they spend interacting with or complimenting teammates)
- Increase in the times people positively challenge one another to give their best

All of these are fantastic indicators that the team is firing on all cylinders. Make sure that your team knows that you have taken notice of their high level of performance and teamwork, and encourage them to do more of the same!

16

Stress and Anger Management (an External Perspective)

No, this chapter isn't devoted to the fine arts of deep breathing, meditation, yoga, or aroma therapy. But it is about something that we feel to varying degrees, all day, every day, but rarely acknowledge or talk about directly. That something is *anxiety* and its close cousin, *anger*. The better that you understand the dynamics of anxiety and anger, particularly as they pertain to your relationships with your owners and architects, the more effective you will be at minimizing the detrimental disruptions caused by these powerful emotions.

When most of us hear the word *anxiety*, some sort of disorder comes to mind, and this is certainly the case for some unfortunate individuals. But the reality is that if you are alive, you experience anxiety. It is the state of arousal that stokes our sense of urgency when deadlines loom and fuels our overall drive to do well. It is also the emotion that gives negative reinforcement its kick.

In the past twenty years, neuroscientists have given us a much better glimpse of how our brains actually work in relation to our emotions. While it has been understood since ancient times that we have both rational and emotionally reactive sides to our personalities, until recently, they were always viewed as separate and disparate functions—one having little to do with the other. But current research has proved this dualistic view false. As Daniel Goleman, the author of *Emotional Intelligence*, asserts,

> The lopsided scientific vision of an emotionally flat mental life—which has guided the last eighty years of research on intelligence—is gradually changing as psychology has begun to recognize the essential role of feeling in thinking. (2005, 41)

To grossly oversimplify, we literally have three brains, layered one on top of the other; the rational, "thinking" brain—the neocortext—where complex functions (such as interpreting blueprints) are processed, and the brainstem, which regulates breathing, circulation, and other automatic responses, (such as our fight-or-flight response). They are connected via the limbic system, which includes the prefrontal lobes and the amygdala, and serves as the seat of our emotions. So, rather than being separate, our brains are hardwired, particularly in the face of a perceived threat, to flip back and forth between our thinking and reactive brains—often in just milliseconds—and all of this is registered in the form of memories by our emotional brain. Our reactive brain takes over to get us out of immediate danger; our rational brain processes pertinent information in order to help us avoid similar situations in the future, and our emotional brain stores the resulting feelings so that we can become mobilized to take action at a moment's notice. As Goleman states,

> When an emotion triggers, within moments the prefrontal lobes perform what amounts to a risk/benefit ratio of myriad possible reactions, and bet that one of them is best. For animals, when to attack, when to run. And for humans … when to attack, when to run—and also, when to placate, persuade, seek sympathy, stonewall, provoke guilt, whine, put on a façade of bravado, be contemptuous—and so on, through the whole repertoire of emotional wiles. (2005, 25)

As much as we might like to believe that our actions are guided by pure reason alone, this really isn't the case. The truth is that our rational brains are often hijacked by our limbic system. This is why, when reason has been restored and we look back on ourselves after "losing it" during a flood of emotion, we often feel like we were a totally different person. In many respects, we were. This relationship between the rational and the reactive mind is even more complicated when we look at how we arrive at decisions. Antonio Damasio (1994), a neurology professor at the University of Iowa, has studied the relationship of emotion and decision making in an unusual set of patients—those with lesions in the amygdala, an area in the brain specifically responsible for processing and remembering emotions. Though intellectually intact (as measured by IQ testing pre- and post lesion), their lives quickly began to unravel. Within a year of diagnosis, many of these patients often lost their jobs or their marriages fell apart. What was the reason behind the unraveling?

Though their intellect was unchanged, they now lacked the ability to process emotions effectively, and as a result, their decision-making ability suffered. As it turns out, Descartes had it wrong; it is not "I think, therefore I am." Well-reasoned decisions are dependent on our ability to process emotions effectively. In a very true sense we have to feel our way through decisions in order to make good ones. So, it's more like "I think *and* feel, therefore I am." Again, reason and emotion are closely tied. What Damasio's patients experienced was a severing of these ties; since they could no longer anticipate the emotional impacts of their decisions, they usually made poor ones. For instance, since they could no longer anticipate feeling badly should they arrive late to a meeting or fail to live up to a deadline, they did both with alarming frequency. And since their ability to edit comments was regulated by this same system, both their work and personal lives suffered as they tended to blurt out any hurtful thing that came to mind. It wasn't long before their bosses and spouses lost patience with their thoughtless behavior and severed some ties of their own.

So, what, precisely, does any of this have to do with the job site and our relationships with our external partners? Well, just about everything.

For example: Let's say an electrical contractor has just been awarded a contract on a multiphase public school project that is scheduled over the course of a ten-year span. Needless to say, they are pleased about having been awarded the work, but aren't all that excited—after all, it's not the high-end, fast-track, high-tech, tool install type job that truly gets their juices flowing. Given the amount of time they have to plan, their pace of getting up to speed on the project could be characterized as lacking a sense of urgency. And, physiologically speaking, there would be some real truth to this assessment. Simply put, when we experience too little anxiety we're generally not aroused enough to fully engage in new tasks. For the same reason that we put off working on that term paper until December, even though it was assigned in September, when we perceive that we have all the time in the world, we simply lack the emotional impetus to move forward with focus and passion.

For their part, during this same time period, the owner has received updated cost projections from the general contractor (GC), based on their revised internal estimates—and the numbers aren't pretty—particularly on the electrical side. This prompts political factions within the ownership group—those who had been skeptical about the project from the start—to make some serious noise about potential budget overruns. In so doing, they cast serious doubt as to whether their managing partners can exert sufficient control to keep costs down. As a result, the anxiety levels for the owners in charge of overseeing the project are now running dangerously high.

Now, let's fast forward two weeks to the first serious meeting between the owner, the GC, the architect, and the primary subcontractors. On the owner's side, the tension is palpable, while a glance over to where the electrical sub's people are sitting reveals broad smiles all around. So, is it any wonder, then, that when the owners get wind that the electrical contractor has barely cracked open the drawing that they explode, while, for their part, the folks from the electrical contractor come away from the meeting believing the owners to be maniacal lunatics? Given their contrasting emotional priming, how could the exchange have gone any other way? After all, for the past two weeks the electrical subcontractor's people have been living comfortably in their neocortext, while the owners have been smoldering away in their overcharged limbic systems. Because the subcontractor can't feel the emotions of the owners, they experience their explosion as irrational and out of control. Similarly, since the owners can't feel the emotions of the subcontractor, they misinterpret their laid-back behavior as cavalier, wasteful, and disrespectful.

So, how can we regain our balance in such emotionally mismatched situations? First is by understanding some basics about anxiety and its influence on performance. The second is to better understand the relationship between anxiety and anger, and to utilize the same principles that helped build our internal teams with our external partners.

ANXIETY AND PERFORMANCE

The relationship of performance to anxiety forms an almost perfect bell curve (Figure 16.1). We perform best when experiencing a moderate level of anxiety. If we experience too little anxiety (low arousal), we often perform well below our capabilities. But if we experience too much anxiety, we "flood out" emotionally, and also perform poorly.

This over/under pattern of responding to anxiety is directly tied to the internal survival mechanism that is preprogrammed in all of us. Believe it or not, 85% of our brain is hardwired for the fight-or-flight responses. That means that our brains are specifically designed to take in data from our

Anxiety and Its Relationship to Performance

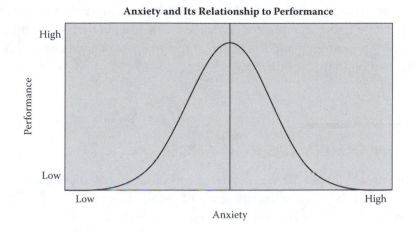

FIGURE 16.1
Anxiety and its relationship to performance.

sensory organs—our eyes, our ears, our sense of smell, our sense of touch—and if danger is detected, translate it to immediate action. So why do we exhibit this over/under response pattern? In short, because it aids our survival; it is a vestige from our tribal days when lions—not deadlines—were our greatest worries. Fight or flight works like an on/off switch, triggering the secretion of hormones and blood sugars that get our hearts pumping, our muscles moving, and our brains scanning for escape roots. A highly charged arousal response is perfect if you have to get away from a lion, but not so great for long-term stressors where reason, logic, and planning are required. Conversely, when our senses don't detect a threat (or the lion gets somebody else), our systems shut down in order to conserve energy for when it is needed in the future. Therefore, our lack of urgency in the absence of danger is not laziness—it is actually survival enhancing. Again, this is great for living in a jungle, but not so great when working at a modern job site.

The question then becomes, what is the optimum level of anxiety for people to experience and still perform well? The answer is somewhere in the middle of these extreme states, where arousal is sufficient enough to trigger engagement, but not so much that emotion takes over and completely hijacks reason. Unfortunately, establishing this moderate ideal state is difficult, as it varies from individual to individual. Some people are knocked off kilter by the slightest criticism, while others require a smack upside the head with a 2 × 4 just to get their attention. Even within individuals, the amount varies. I've known burly, highly confident superintendents who can handle tightly compressed schedules with ease, but who are quickly reduced to sputtering puddles of goo at the prospect of giving a two-hour presentation in front of their peers.

WHEN ANXIETY TIPS INTO ANGER

As Ben Franklin put it: "Anger is never without reason, but seldom a good one." Anger is triggered by the sense of being endangered. But as Dolf Zillmann, a psychologist at the University of Alabama, notes,

> In humans, endangerment can be signaled not just by outright physical threat, but also, as is more often the case, by a symbolic threat to self-esteem or dignity: being treated unjustly or rudely, being insulted or demeaned, being frustrated in pursuing an important goal. (Goleman, 2005, 60)

This, then, is the critical component for when anxiety tips over into anger. Not only is the limbic system fully charged for action, but it also hijacks the neocortex into giving justifications for our aggressive impulses, based on a perceived threat. As Goleman notes,

> The amygdala may well be the source of the sudden spark of rage.... But the other end of the emotional circuitry, the neocortext, most likely foments more calculated angers, such as cool-headed revenge or outrage at unfairness or injustice. (2005, 59)

What makes all of this more difficult is that, unlike sadness, anger is energizing. In the short run, it feels good to be in its grasp and to do battle. We even feel justified in doing so. But it is the midst of such takeovers, Goleman argues, that we need to increase our emotional intelligence. Despite our biology, we, at the jobsite need to understand how, in the long run, ongoing battles, fueled by emotion, can lead to disruptions, waste, lost productivity, and lawsuits.

So what can you do when the heat of the moment threatens to overtake what could otherwise be a solid working relationship?

Much of what you can do to defuse such volatile situations centers on the same principles that you employed with your internal team: empathetic listening, extending invitations of trust and commitment, and exhibiting the willingness to engage in healthy conflict. The trick is to identify when a state of productive arousal is about to tip into nonproductive high anxiety early on, and engage in countermeasures to help move it back to a more productive zone.

As described in the continuous assessment chapter, a predictable tell for an owner is when they begin to scrutinize your decisions, work plans, or schedule. They will often display their rising anxiety by becoming, sarcastic, belligerent, or controlling—or by insisting that every prescribed policy and procedure as outlined in the contract be followed to the letter of the law. As a last resort, they will threaten to zip up their wallets until they are indeed heard.

For their part, while being some of the most brilliant and creative professionals on the planet, when highly anxious and teetering on anger, architects often choose to become conflict avoidant and passive-aggressive versus openly combative. I've witnessed the following with a fair degree of frequency (extracted from one of my own assessment reports):

> Rather than phoning the GC's PM and telling him directly that they were displeased with his team's vetting of requests for information (RFIs), the architect simply stopped responding to RFIs, assuming that this would be sufficient to convey his displeasure. Rather than getting the message, the GC viewed the A&E as withholding vital information and unnecessarily delaying workflow; thus, the war between these two entities began.

If your external partners display any of the behaviors listed, please resist the urge to interpret this as confirmatory evidence that they are unworthy of your empathy. They are simply human beings acting in ways that are all too human. Take a step back and try to gain a little perspective.

UNDERSTANDING THE OWNER'S PERSPECTIVE

While owners can appear demanding, unrealistic, and often seem to want the impossible accomplished overnight for no additional cost, it is important to look at things from their perspective. What would be important to you if you were in their shoes? Certainly, you would want to know that you were getting value for what you are paying for. In fact, this is usually at the core of an owner's greatest anxiety, that is, that project costs will keep escalating, and in the end, they won't get anything close to the building that they were hoping for. Also, keep in mind, like you, owners are accountable to someone else—be it a board of directors, stockholders, regulatory agencies, or the bank that is holding their loan. Therefore, such things as escalating costs or slipping schedules, in their eyes, become very real threats to their existence. So, how do owners usually attempt to manage their anxiety in the face of such threats? Like anyone else: by trying to get rid of it by gaining control. They want information! They demand transparency! They insist upon meeting after meeting! They scrutinize! As their pitch rises to a frequency that only dolphins can hear, it is important that you try to defuse the situation by demonstrating your willingness to hear their concerns and empathize with their plight.

JOIN THEIR ANXIETY VERSUS RESISTING IT

The last thing you want to do in a situation where the owner's anxiety is running high is to dismiss it, dig your heals in, or become countercombative. In these moments, it is important to remind yourself that your goal is not to wage war, but to move the situation from a state of emotional hijacking to the place where reason and logic can once again prevail. Here is a little secret that you can take to the bank: the more assured that the owner is that you and your team have their best interest at heart, the more likely it is that their anxiety will reduce to levels where they can actually hear your plan. So, rather than dismissing their concerns with the perfunctory "We've got it covered," step up and demonstrate *that you actually do* have it covered. Repeat back their concerns until they no longer feel like they have to fight to get heard. Then make sure to use one of the primary tools at your disposal to fully allay their fears—a fully integrated schedule. But to successfully implement this tool, you must go one step further. Owner anxiety is often triggered when they ask how a specific activity is tied to the schedule, and instead of hearing back a specific answer, they are told, "Gee, that's a good question." This leads them to believe that you don't really have your arms around the project, and their fight-or-flight response becomes fully armed. Make sure that your team is able to answer any and all questions pertaining to their area of responsibility per the schedule. For example:

> Let me make sure that I am hearing you correctly; you are concerned about window installation interrupting the overall schedule—is that right? Here is what we are doing about it. As you can see on the schedule, window installation will begin June 8. Our lead engineer identified early on that there was a long lead time for this item, which is why she pushed for RFIs and a decision on this in September, which we got. The buy-out has been complete for some time, and barring any unforeseen circumstances, such as high seas piracy or the ship getting held up in customs for an inordinately long period of time, they are scheduled to arrive on site in four weeks—two weeks prior to installation.

This is the kind of information that the owner *needs* to hear. It is important that you *not* interpret their need to know as an affront to your professionalism. It's just their anxiety overtaking them. But it won't go away until you satisfy this need. And yes, to anticipate your question, some owners are, indeed, needier than others; that's just how it is—people are wired differently.

Of course, this presupposes that your team is actually working off of an integrated schedule. Don't laugh; there are plenty of project teams that fail to produce a meaningful schedule and then are incredulous when the owner goes ballistic. I'm generally not terribly sympathetic to their plight.

GO THE EXTRA MILE TO FULLY UNDERSTAND THE OWNER'S NEEDS AND CONCERNS

At the outset, make a point of asking the owners about what is most important to them and what they are most worried about. Don't assume that the answer is always the same, that is, finishing the building on time and on budget. During a partnering session, I asked the GC to state what they thought was most important to the owner. They gave the standard "on time, on budget, to design" answer. When the owner was asked whether the GC had hit the mark, to everyone's surprise, he frowned. "To be perfectly honest, you could build this building ahead of schedule and below cost and not give us what we want at all." As jaws dropped around the table, he continued: "You need to understand what this building means to us. We're a biotech campus at a public university. We're competing for scientists with top flight companies from all around the world. This will be the first building that they will see on campus. If it isn't flawless; if it doesn't wow them right away and make them want to work here, then this building, in our eyes, will be a failure." When asked what he believed would ensure a good outcome, he said, "I'll be blunt. We spent a lot of time and money vetting the design. We don't want this building value engineered or built on the cheap so you can maximize your savings participation. We want quality. We want it built precisely to specifications so that the wow factor will remain intact." Good to know.

This type of information would have come in handy for a team that lost out on the second phase of a project they were in line to get, even though the first phase finished successfully—at least in terms of the usual on-time, on-budget criteria. When their business development folks finally got around to asking why the owner decided to go in a different direction, they were quite succinct.

> Frankly, we were frustrated with your team. We told them, up front, how important it was for us to be kept in the loop—how much our end users needed to know what was going on and have input. But every time we asked questions, the typical response from your team was, "We'll have to get back to you on that." That in itself wouldn't have been a problem—we expect people to check their facts. The problem was they never did get back to us. In fact, there were times when your team seemed downright annoyed by anything that we wanted to know. We simply don't accept that. So we're going with another GC. They may not have your resume, but we've gained a strong sense that they'll do a better job of keeping us informed and will be far more collaborative in their approach.

Ouch. Just goes to show that it's not always about price for some owners. It also demonstrates that anger can take many forms, including revenge. What better way to get even with a company that you believe has ignored your needs than to not award them the additional work that they thought was coming their way?

Once you have taken the time to find out what is important to the owner, write it down, laminate it, and stick it on a wall where everyone can see it. This is the easiest way to keep the owner's concerns prescient in your team's collective unconscious. Plus, whenever the owner visits the trailer, he or she will appreciate seeing that you've taken the time to raise your team's awareness about what is important to him or her, particularly if this list is updated on a regular basis.

OWNER ANXIETY AS A MATTER OF TIMING

Lou Brugantti in his course "Building Excellence" observed that owner anxiety is usually at its highest when general contractor mobilization is at its lowest points—at the very beginning and very end of the project. It is at its lowest levels during the middle phase, when the GC is fully mobilized. And this makes perfect emotional sense (Figure 16.2).

At the beginning of the job, the owner is feeling the full impact of real costs versus estimated costs and, regardless of their sophistication, will experience a huge spike in their anxiety levels. Conversely, at the end of the job, when additional costs are coming to light, or the end date appears to be slipping, the GC is often in the process of shifting some of their personnel out to other projects. At both ends of the scale, the owner feels as if they are going it alone, and their anxiety peaks accordingly.

This helps explains the extreme volatility, and shear folly, of waiting until the very end of a job to deliver bad news. This is the surest way of leaving a sour taste in the owner's mouth, regardless of the eventual outcome of the project. It is the one thing that will stick in his or her minds forever and will greatly jeopardize landing any future work with them! If you anticipate bad news, speak of it when owner anxiety is at its lowest point—when you are fully mobilized—and well before options to fix the problem have become extremely limited and much more expensive.

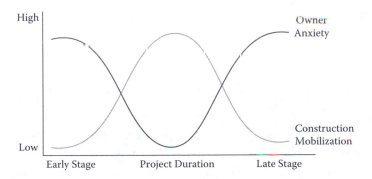

FIGURE 16.2
"Build excellence" owner anxiety analysis.

ESTABLISH CLEAR EXPECTATIONS

Know your scopes. In the spirit of providing great customer service, I've seen many a GC and subcontractor overstep their boundaries and go well beyond what was outlined in their contract. As a result, their ability to execute the work that was in their contract became compromised. Besides doing work that you are not being compensated for, this overreaching creates another problem; it establishes a false baseline of expectations in the owner's mind. These additional services that you are "throwing in" as an act of good faith will now be expected of you and your team throughout the duration of the project. Later on, if you try to pull back your performance in line with the original parameters of the contract, this will be viewed as a takeaway by the owner, and you'll likely be accused of no longer doing your job. This is a complete misperception, but as it is fueled by anxiety, no amount of explanation will alter it. There is a fine line between being accommodating and doing too much. Anything that impinges upon your resources to the point where you can't do the job you were hired to do should be addressed early on. For instance:

> We can do the additional policing and reporting for the safety program that you are now requesting, but it is well outside of our scope of work, and well beyond the industry standard. We will require additional staff resources to accommodate it. If you are willing to pay for this, we would be more than happy to accommodate you and provide these services. Otherwise, we will need to stick to our original reporting agreement as outlined in the contract, or our staff will be stretched too thin on their other tasks and this will negatively impact the project.

KNOWING WHEN TO SET LIMITS

At this point in the discussion, you've probably come away with the impression that I believe all conflicts that arise between owners and contractors are entirely due to inattentiveness or insensitivity on the part of the contractor. I know all too well that this certainly is not the case. I have personally witnessed absolutely appalling behavior on the part of owners and owners' representatives, which seemed to arise purely out of the belief that, because they write the checks, they are entitled to treat others with utter contempt and disrespect. In such extreme circumstances, waging war for the sake of what is good and right is not only justified but also necessary. But this must come from a place of measured reason versus a gut-level limbic storm. Let me give you an example.

An owner's representative for a college campus took much self-admitted pride in making the lives of contractors miserable. He often bragged about the number subcontractors that he had put out of business over the years, and thought himself clever and gifted at "being able to beat them at their own game." The reality was that he was neither clever nor gifted; he was simply skilled at being a bully. He often used the most appalling language and relied on making outrageous and humiliating accusations in order to get his way. Sadly, his behavior was considered effective by many of those who hired him (oblivious to the "asshole tax" that many companies included in their bids when they found out who the owner's rep would be). At one point, his behavior became so extreme (he cursed out a young female engineer publicly for speaking out of turn) that the PM for the GC decided that he had had enough. At the next owner/GC meeting, the PM set a tape recorder on the table in front of him. When asked why, he didn't hesitate. "We're here to do a job, not to be verbally abused. I'm open to hearing your concerns, and if you personally have a bone to pick with my staff, I am all ears. But I'm not subjecting my staff to Mr. X's verbal tirades and wild accusations any longer. If it continues, I'm going to assist my employees in pursuing a claim of creating a hostile work environment against him." It should be noted that the PM did this with full knowledge and endorsement of his top management—and the blessing of the owner—who had grown weary of the owner's rep's tirades, but was too much of a weenie to confront him directly. Whether or not such a tactic would have stood up in court is debatable, but the PM was successful in getting the owner's rep removed to a behind-the-scenes role, thus shielding his staff from future unwarranted barrages.

SLOWING DOWN THE RAGE TRAIN

In most instances, when owners behaves badly, it is because their fight-or-flight response has been tweaked and they are having a hard time articulating the thinking behind their emotion. If we choose to respond in kind, their anger will build into rage and, like an out-of-control freight train, will threaten to take out everything in its path. Instead, try to give them the benefit of the doubt and slow down the train by demonstrating your willingness to explore the root cause of their angst:

> I can see that you are upset, and I take that very seriously. But to be completely honest with you, I'm not entirely clear on what you are specifically upset about. Can we pump the brakes, take the issues one at a time, and sort this out so we can get to the bottom of what is causing the problem?

If you've tried everything you can think of to defuse the situation, and the behaviors (typically screaming and yelling) continue, it's perfectly acceptable to pack up your stuff and say:

> We're obviously not going to get anywhere right now. We are all much too charged up. I know I am. Personally, I'd like us all to take the opportunity to calm down so we can get to the root of the problems rather than continue to tear each other apart. It's 11:00 a.m. now. Could we try this again after lunch—say, 1:00 p.m.?

You'll be surprised at how often this method works. Despite their protests to the contrary, most people in construction don't enjoy it when things turn ugly. But their brains are so hijacked by emotion that they feel doing anything other than continuing to do battle is somehow akin to losing, so they won't disengage. In truth, most people are relieved when somebody has the guts to call a time-out so that everyone can regain their composure.

Here are some additional tools that you can use:

- Don't respond to accusatory emails in kind. If you do write a reply, wait twenty-four hours before hitting "send" and edit accordingly.
- Restore your sense of purpose. Remind yourself that you are not on site to win battles, but to help someone else's dream become a reality—for all the good and the bad that this entails.
- Correct your own funky thinking. If you catch yourself thinking, "Who does he think he is?" "I'm nobody's punching bag!" or "They think I'm a lying jerk? I'll show them what a lying jerk really is!" it is time to recognize that your own fight-or-flight response has become fully armed and to invoke a self-imposed moratorium on taking action on these thoughts.

- Collect your composure. You are human; you will become angry. Give yourself some time to talk yourself down off the ledge. Vent to someone you trust, but when doing so, don't build a case in your head that justifies retaliatory behavior. Keep working toward the *best* solutions.

UNDERSTANDING THE CONCERNS OF THE ARCHITECTS

I've heard every criticism about architects under the sun, including

- They don't care one wit about constructability; all they care about is aesthetics.
- They could care less about the cost; they just want the building to look pretty.
- They are unresponsive; they don't give a damn about the impacts that their delayed responses cause.

While, on a case-by-case basis, there may be some truth to these accusations, collectively, architects, in my experience, are often convenient scapegoats for general contractors and contractors, while their own needs and concerns are usually the least taken into account.

Again, try to walk in their shoes. If you were them, what would you care about? Let's face it: most architects care a lot about what other people think. They have to. Their aesthetic sense isn't just a point of pride; their future depends on it. When all is said and done, no one in the community will care that the building was value engineered so the owner could save a ton of money and the contractors could increase their profits. What will count is how it looks and whether or not the building fits within its surrounding. By and large, a building is judged entirely on its aesthetic qualities. And if it looks awful or out of place, this albatross won't be hung around the GC or subcontractor's neck; it will fall squarely on the architect—and will in turn negatively influence future sales. So, can you blame them for balking at your cost-cutting ideas that downgrade the building's aesthetics to an ugly, all-purpose box?

There is also a pragmatic reason for their obstinacy. While the GC and subcontractors are on the hook for latent defects for a legally specified duration (ten years in California), the A&E is on the hook for life. That's why they are so rigid about their engineering and design criteria. They can be sued in perpetuity if, at any point, the building fails.

THE DESIGN AS AN EXPRESSION OF VALUES

But there is another issue at play, and it has to do with values. It is a complete falsity that the A&Es don't care about the budget. They do. But to them, their design symbolizes something very important. It is a statement that they listened carefully to what the owner said they wanted. To the architect, their design, in a very real sense, is their promise to fulfill the owner's dream. So your efforts to alter the design, which to you may seem well intended or, at the very least, benign, won't feel that way to the architect. They will feel as though they are being asked to break the promise to the owner. And anyone who cares about integrity will experience a spike in his or her anxiety, and behave badly, when placed in such a position of compromise.

Also, the A&Es want to be viewed as an integral part of the decision-making and problem-solving process throughout the life of the project; they very much want to be included in budget and scheduling issues versus being viewed as an impediment to the building process. They don't like being pushed aside or relegated to the role of mere RFI responders. Of course, complicating this is the fact that much of the architect's fee is burnt up during design development, thus giving them the appearance of being disinterested, when in fact they are merely trying to conserve their own finite resources.

When architects are angry, it won't appear as overt as the owner's response. Years of experiance of walking a tight rope between owners and the GCs have shaped their behavior, so they tend to respond passive-aggressively—choosing to resist rather than becoming overtly aggressive so as not to depolarize either relationship. The difficulty when experiencing such behavior is to not lash out or respond in kind. You will have to work hard to remind yourself that the goal is to keep them actively engaged versus treating them like a necessary evil. Therefore,

- Invite the architects into constructability and budgetary discussions.
- Ask for *their* cost savings ideas, that is, changes that *won't* compromise the overall design, but will still save the owner money.
- Do mock-ups to get subcontractors excited about the design so that they will *want* to honor it.
- Ask the architects *directly* why they have stopped responding to RFIs or are otherwise appearing to resist the building process

- Invite their vulnerability and seek to understand their point of view. (Have they stopped responding to RFIs because they are out of money? Short on staff? Upset about insufficient vetting of RFIs?)
- When they take the risk and are vulnerable, don't use it against them. (If they confess to being out of money, join with them to find ways to make economical use of their time, such as full-day design reviews of particular trades versus submitting issues one at a time.)

LEARNING TO OPENLY EXPRESS YOUR OWN ANXIETY

When dealing with your external partners, take the risk of being the initiator of vulnerability. Make sure that you take the time to explain to the owner and A&E what is important to you. Let them know your trigger points and the things that your upper management expects you to execute and be accountable for. If you're a subcontractor, remind them that your greatest risk is in manpower costs, which is why you keep harping on coordination and scheduling issues so loudly. If you're a GC, explain why prompt payment and timely decisions are so vital for keeping the assembly line moving, which in the long run saves everyone time and money. But most of all, remember the law of reciprocity! If you want the owner and A&E to actually care about your concerns, you are going to need to demonstrate that you care about theirs!

PARTNERING

I conduct partnering sessions, and I am a big believer in the process, provided that it is grounded in reality versus hyperbole. Instead of wasting time creating mission statements that nobody reads or follows, establish rules of engagement that acknowledge the ugly emotional and behavioral realities that can creep into any building process. Here is an example:

> We, the undersigned, agree to do the following to the best of our abilities:
>
> 1. We will keep our goals in the center of our focus (no private agendas).
> 2. We will memorialize our agreements after we reach a verbal agreement (not as a "gotcha" but in order to establish and clarify "the rules of the game").

3. We will develop a mindset that at the end of the day, it's all about the people and our relationships.
4. We will elevate issues early on, while they are still fixable (and less expensive).
5. When we need to elevate an issue, we will do so jointly and with clear expectations (i.e., "Can we agree that we'll give this a week, but if we can't resolve this between us by then, we will elevate it together?")
6. We will coordinate our actions. We will interact to ensure that things go right, rather than working in silos, and fixing things that have gone wrong.
7. When we have discussions, we will ask ourselves "Who else needs to know this?" and work to include them.
8. If we are accidentally left out of a discussion, we will give our partners the benefit of the doubt that it was not their intention to do so and will ask for our seat at the table of ideas.
9. We will be truthful about our goals and needs (don't cry "urgent" when this isn't the case).
10. We will encourage passionate debate as a means of obtaining commitment. We will discuss issues passionately, but will not let issues become personal or linger because we are trying to avoid a conflict.
11. When we are confused and need clarification, we will ask for it directly (rather than sulking or complaining behind one another's backs).
12. We will listen to the needs and concerns of others and endeavor to treat them with as much care as we do our own.
13. We will strive to model and foster vulnerability. We will admit when we are having problems, don't understand something, or have made a mistake. We will honor others when they exhibit vulnerability (rather than using it against them).

17

Generational Issues

I was reluctant to include this chapter, but since it is such a hot topic among senior managers no matter where I travel, I felt compelled to discuss it. Their complaints usually sound something like this: "These kids today don't want to do what it takes to get the job done, and they don't respect those of us who do. They completely lack initiative and need constant direction. If I'm not there to tell them what to do, things just don't get done. Besides that, even though they have very little experience, they think they should be promoted in a heartbeat. Hell, I've got one kid on my job who thinks that by next year he should be doing *my* job. He doesn't have a clue about what he doesn't know."

Before I comment, however briefly, I should disclose that I, myself, am an old geezer so I'm, therefore, inclined to hold a few biases of my own. But I don't think the generational gap that we are experiencing today is beyond the norm. Tension between generations is age old. As Plato decried some 2,400 years ago:

> What is happening to our young people? They disrespect their elders; they disobey their parents. They riot in the streets influenced with wild notions. Their morals are decaying. What will become of them?

Even prior to WWII, many wondered aloud if "this selfish and soft" generation, now known as the "greatest generation" would have the requisite metal to fight a common enemy.

I'll admit that over the years I have met a few young people that have caused me pause. But I have also met scores more who are more than willing to make the necessary sacrifices to get a job done. The Seattle football stadium project cited earlier was heavily populated by twenty-somethings, yet most worked well into the evenings, including Fridays and weekends, doing whatever it took to do a quality job. Conversely, I've met a significant number of older workers who, while talking a good game, did little to justify the high price tag that they commanded. Certainly, there are differences between the generations, but they are more a matter of degree and, as has always been the case, are dictated by the cultures in which they are raised.

In the past twenty years there has been enormous pressure on young people to get along socially and to achieve. So from the time most of them could walk they were carted off to ballet, t-ball, hockey, soccer, karate and various other activities. An interesting artifact of all this is that the vast majority of these activities were organized, led, and refereed by adults. In short, this generation is used to being heavily controlled and governed by others. Add cleaved marriages into the mix (my niece was the only child in her San Francisco high school homeroom class whose parents *weren't* divorced), and all the schlepping from place to place this entails, and the strong demand for this generation to take direction, adjust, and simply "get along" truly comes to the fore.

I don't know about you, but when I was growing up, my and my friends parents told us when to be home for lunch and supper—and that was about it. All of that time in between was left to us to figure out. Our parents didn't know much about what we were up to and, I think, as long as the authorities weren't involved, preferred it that way.

So what were we up to? Besides secretly hoping that our parents *would* get a divorce, we spent much of it coming up with games to keep ourselves amused. Due to less than optimum resources, we often had to devise modified versions of football, baseball, and basketball, make up our own rules, referee our own games (was there anything purer than the "do over" rule when things became hopelessly deadlocked?), and generate methods to regulate our behavior amid a rather fiercely competitive environment (what's all this nonsense about *not* keeping score?).

All of this is very different from how kids are raised today. To be honest, I'm not sure which way is better—being under- or oversupervised—as both clearly have their pluses and minuses. But what is certain is that each condition produces different normative values for each generation.

Not surprisingly then, people in our generation are great at competing, working independently, taking the initiative, and figuring things out on our own. And these are the very qualities that we have come to value. Conversely, we're not always so great at sharing credit, working as a cooperative team, or communicating our concerns in a productive way. We are much more prone to fly off the handle and mend our fences later, rather than discussing things in a calm manner. But on the flip side, we're usually pretty good at not glossing over problems or concerns.

Today's generation tends to get along well with others socially, has little trouble sharing information and working cooperatively, and is fairly good at expressing their feelings and concerns in a productive manner (though, at times, they can be a tad conflict avoidant). While often not well schooled in such basics as grammar and letter writing, they are incredibly comfortable and proficient with new technology. They also have a wide range of interests that tend to make them a bit more balanced than their workaholic counterparts. And these are the qualities that *they* have learned to value.

Another key difference is in the area of affluence. Many children of today's generation were raised in an environment of unprecedented material excess, and rarely were put in the position of having to make do. It's probably not surprising then that some in today's generation believe that the world revolves around them, because between all the activities that they were chauffeured to and from, and the affluence bestowed upon them, it pretty much did.

So what happens when these same young people hit the workforce and encounter bosses comprised of those from an older generation? Since each grew up with a different notion of what is valued, each comes primed with entirely different sets of expectations for what is acceptable behavior in the workplace. Many older workers believe strongly in self-sufficiency and initiative, yet given their conditioning, many young people struggle with these same attributes. And since some young people were raised in a manner where self-centeredness and merely getting along were prized, they often are shocked or become extremely hostile when given feedback that their performance does not match expectations.

So what's a manager with some gray around his temples to do? First off, lighten up a bit. The young people on your projects aren't an alien race. Do you remember the adage, "Don't trust anyone over thirty"? That wasn't the brightest philosophy in the world, but many of us believed it. Though they are not always what we want them to be, they have considerable strengths—you just need to tap into them in a little different way.

Instead of barking out edicts and expecting your young charges to fill in the blanks, this generation grasps information more effectively when they have the chance to explore it in a social setting. For instance, let's say that you are trying to get your staff to pay attention to the particulars of the contract. Rather than commanding them to go read it, host a lunch where you hand out ambiguous portions of your contract and lead them in a spirited discussion about the vagaries of contract interpretation. This will enliven what can be a fairly dull topic anyway (yes, this generation does like to be entertained), and teach them some important and humbling lessons about the intricacies of construction law.

If you notice that your young people are stumbling over the same obstacles, or if you are about to enter into a new phase of the work that they have never done before, conduct an impromptu group training session. The topics can include how to read a schedule, how to write a potential change order (PCO), how to close out a subcontractor, etc. If you work in a large company, don't hesitate to ask experts from ancillary departments to conduct a brief training session, in, say, cost or accounting methods.

Also, take the time to teach the broader context of the task, for example, *why* well-written PCOs, RFIs, and submittals are of such vital importance and *how* they fit in the larger scheme of the project. Also have high-quality procedural examples at the ready to serve as a resource so people will have an idea of the target that they will need to hit. And if you want to teach certain behaviors, have real-life examples from which to draw. For example, this is why Janice Smith is considered such a good engineer: She reads the contract, thoroughly reviews all documents and drawings, and treats all her subcontractors with respect. But she still expects them to verify that they are actually doing what they said they were going to do by going out to their prefab yard and seeing, with her own eyes, what thay have assembled.

I know what some of you are thinking, "Nobody did that for me—I had to figure things out for myself!" Well, get over it. Given the complexity of tasks that we ask them to do and the sheer volume that our young people are now expected to handle straight out of school (do you remember the good old days when rookies only had to track one trade with a fairly low dollar amount?), I don't think making people figure things out on their own is such a wise strategy.

That's not to say that you always have to hold their hands or give them the answer. When warranted, I strongly encourage you to push your young staff to learn how to think for themselves. For instance, if the nature of their question is contractual, I recommend the following: "I could give you the answer, but I know it's in your contract. Go research it. Let's meet in two hours and then you can tell me what you think the right answer is."

When I cited this example in a training class, a young engineer shared his own experience:

> My old project manager used to do that very thing with me all the time. To be honest, I hated him for it! I thought he was just being an a—hole for not answering my questions, and that he just enjoyed messing with me. But a year later, I got assigned to another job. And it was weird. Not only did I now know where to find the answers to my own questions, but I was actually helping the new people do the same. It was then that that I realized what he had been doing. Just the other day I called him up to thank him. I hated him then; now I really couldn't appreciate him more for what he did.

For that small, but thorny minority who are indeed a little too full of themselves, I recommend a little different approach. Let them fail, but in a controlled way. Let them run a meeting that they think they are ready to run, but aren't. With the project executive's or division manager's blessing, let them generate a cost report, and allow the project executive (PX) or division manager (DM) (with a coordinated heads up) to (professionally) rip them apart. This isn't being manipulative if the true intent is to help them mature. Nothing schools arrogance and self-centeredness more than an extra helping of humble pie. The quality people amid this crowd won't like it, but they will be the better for it. Those who can't handle it will quit in a huff. And that's okay. Perhaps some of them will return down the road a little wiser.

Despite our differences, this generation needs our help as much as we needed help from the generation that preceded ours. Through you, they will to learn how to better serve the needs of their teammates and their customers. Who knows, maybe someday one of them will call you up and thank you for it!

18

Personality Testing—Don't Do It! (Better Ways to Know and Understand Your Staff)

More often than I'd like, I am presented with a glossy brochure by a hopeful VP asking my professional opinion about the latest and greatest personality test that he is thinking of administering to his staff. And invariably, he is disappointed by my response—that most personality tests don't deliver as advertised.

The consultants who tout these tests often extoll their scientific mend, claiming that, based on responses to a brief questionnaire and by scanning a graph or color-coded chart, you'll be able to see whether or not you and your people are introverted or extroverted, play well together under stress, or will perform comfortably at presentations. Some even assert that these tests can be utilized to determine if people are well suited for their present positions or if they are good candidates for promotion.

Don't be fooled by the glossy packages and the authoritative presentations by these purported experts. Like most things that sound too good to be true, most of their assertions are far from the truth. Let me briefly explain why you should be extremely skeptical about all of these claims, particularly those regarding their scientific basis, and why your time, energy, and money would be much better spent engaging with your team in more meaningful ways.

WHAT DETERMINES IF SOMETHING IS SCIENTIFIC?

Beginning in the 1940s, the study of human behavior came under much more rigorous scrutiny and began to employ scientific methods in its research paradigms. Simply put, as in the hard sciences, if a theoretical assertion is made, in order for it to be considered scientific, it must be submitted for peer review. Dependent and independent variables must be tested against control groups, and their differences measured. Findings are then subjected to statistical tests in order verify that the results obtained were not simply due to chance. These results are then published and subjected to further peer review.

So how does personality testing measure up against this type of scrutiny? In short, not very well. Many tests have what is called face validity; that is, they appear to measure what they claim. For instance, some of us do become energized in social situations, while others tend to experience the opposite effect and become drained. And at least on the surface, these tests seem to cull out these differences. But are these results scientifically valid and reliable, and do they have predictive value as claimed?

In terms of validity, most people who administer these tests only have one way to measure whether or not their test is accurate: they hand a person a profile and ask, is this true for you? In most cases, people will say yes. But here is the problem: like horoscopes, most personality profiles fit roughly 90% of the people they are handed to. For instance, a finding that says, "You don't easily trust others," would fit just about all but the horrifically naïve. In actuality, this doesn't tell us anything new or profound about the individual in question, other than that the person is normal. What is rarely tested is whether or not, if randomly handed a different profile, the person would also say, "Yes, this fits me."

But it is in the realm of reliability that personality testing really takes a beating. For a personality test to be considered scientific, the results obtained shouldn't vary across a variety of dimensions:

- Mood. The person should score the same regardless of his or her particular mood (i.e., the test shouldn't generate a different profile based on whether or not the person got up on the wrong side of the bed).
- Time. If you are testing for something as enduring as personality, the results should not vary wildly over time. If you test me today, or a year from now, I should produce the same profile.

- Intentional misrepresentation. If a test is indeed scientific, participants should not be able to manipulate the results.

On all counts, when analyzed statistically, the vast majority of personality tests utilized in the workplace today fail the reliability test miserably.
Mood does affect most of these tests. Upon retest, the most commonly
used personality tests produce entirely different profiles nearly 40% of the
time. And on most tests, there is no way to detect manipulation. (The one
exception is the Minnesota Multiphasic Personality Inventory [MMPI],
which is used diagnostically to assess serious mental illnesses such as
bipolar disorder, schizophrenia, paranoia, depression, etc. The designers
of the MMPI anticipated that people would lie to portray themselves in a
particular light—healthier than they are or sicker than they are. Therefore,
a "lie scale" was intentionally built into the test to detect this. Rather than
30 questions, as is standard for most workplace tests, they ask 1,500, thus
having far more data points to analyze and compare.)

Shamefully, the above facts are well known to those peddling these tests.
If you want to stop any of these purveyors of pseudoscience in their tracks,
just ask them about their test-retest reliability coefficients. If they mumble
something about how their test meets all standards, ask them to produce
hard numbers. And if they state that it is not their policy to retest people,
quickly escort them out of your building.

If your company insists on giving you a personality test, I'd recommend
that you go online to http://Buros.unl.edu. You can obtain a copy of a
scientific analysis conducted for most personality tests on behalf of the
Mental Measurements Yearbook (the equivalent of the FDA for psychological testing) by paying a mere $15. After you get the report, pay a little
visit to your human resources department and see if they still think it's a
good idea to administer these tests.

WHAT'S THE HARM?

"Come on, Gary, lighten up. If these tests get people communicating, isn't
that a good thing? Besides, the world is an unpredictable place, and engineers aren't known to be people experts, so what's the harm in them gaining a little insight into the people that they work with?"

Okay, these are fair enough questions, and I hear them all the time. So I'll answer them directly. In my opinion, there is quite a bit of harm that can result from the use and misuse of personality testing in the workplace.

Waste of money and time. If I gave you a list of your subordinates and asked you to rate them on whether they were introverted or extroverted, or thinking or feeling types, couldn't you do this with a high degree of accuracy without giving them a test? More importantly, what does this information do for you? Since, via these tests, you'll be adding an extra layer of complexity by forcing people to learn a new (often made up), jargon-filled language, all you will really be doing is substituting obscurity for knowledge.

Encourages labeling. We have enough problems in the workplace with sticking dismissive tags on people. How many times have you heard people described by others as *lazy, stupid, arrogant, self-righteous, egotistical*, etc.? So what's the difference if we tack on a few more like *analytical, thinking, feeling, introvert*, etc.? Here's the problem with all of this: labels diminish people and get in our way of truly getting to know them. Once we put a tag on someone, we think we know them better than we really do and tend to stop seeing them as people. They no longer have unique likes or dislikes, or possess particular talents and skills—they simply become their label. And at work, once we hit a roadblock and know the label, we simply stop engaging with them. ("I know Gary; his test says that he's an analyzer. I know that I won't get a word in edgewise, so why bother?") Given the poor reliability of these devices, do you really want to rely on someone else's tag to determine how you will relate to someone?

Gives a false sense of predictability. Similarly, once we know the label, for some reason, we think we can predict what that person will do in the future. In short, we put them in a box. The label, in essence, becomes shorthand that says that we need go no further. And this in turn becomes a self-fulfilling prophesy. For instance: "Chris's testing says that he's an introvert, so we probably shouldn't put him on the presentation." "Why didn't we put Chris on the presentation? Because he's an introvert!" Honestly, what is more demeaning than reducing a person down to a small number of tendencies, then limiting his or her future opportunities based on that label? A friend of mine said it best: "I can know that my dinner contains beef, salt, pepper, red wine, paprika, and garlic, but does that really tell me anything about how my dinner is going to taste?"

Gives us an excuse for our own bad behavior. One company I work with uses a test that generates color profiles that supposedly represent various aspects of a person's personality. When I presented my assessment findings, and informed a particular manager that her people felt that she shut down healthy conflict by becoming overly confrontational when she perceived that others were questioning her authority, she laughed and presented her color graph. "Oh, that's me all right," she said. "I go all 'red' and analytical when I'm under stress!"

This manager was actually using the test to justify why she was not listening to her teammates—and thus failing to build trust with them. It was almost as if she had come to believe that since everyone knew her test results (they were actually posted on her desk for all to see), that it was her coworkers' responsibility to accommodate her bad behavior and "approach her correctly," rather than her job to listen to them effectively. Rather than helping her with her work relationships, the test actually got in her way.

Misuse. Despite all the disclaimers to the contrary, people in authority can't seem to resist using these tests in ways that they were never intended. Even though, time and again, personality testing is shown to have little predictive value, such tests are still misused to determine whether or not someone should be hired, promoted, or retained. Some top managers also favor specific profiles over others (usually ones that are similar to their own), even though the poor reliability standards of these tests render comparisons between people completely impossible.

Potential legal nightmare. If the workplace is expected to maintain privacy standards, isn't it a bit odd to parade something as personal as someone's personality profile for all to see? What is possibly more personal and private than that? Plus, if you terminate an employee, and they happen to have been tested recently, how are you going to prove that your action was not a direct result of the test?

Gives lazy, controlling, or self-absorbed managers a false sense of power and security. The fact is that most personality tests reveal far more about the people that love them than they do about those being tested. Think about those in your company who are the biggest proponents of personality testing and allow me to take a stab at pro-filing them. They usually love the sound of their own voice and are annoyed whenever they have to listen to others, aren't they? And rather than being repulsed by the idea of boiling people down to a few adjectives, they relish the false sense of knowledge and power that such reductionism renders. They love the idea of knowing somebody better than that person knows himself or herself. They love being in control. How did I do?

In reality, these tests, or the idea of them, are great for those who don't want to take the time to actually get to know the people they work with. They see personality tests as a substitute for actual engagement, or a "secret key" by which they can control people. Since their interactions with others are few, they are amazed by the seeming accuracy of the tests. If they spent any quality time actually getting to know their staff, they would soon see how limited in scope these tests truly are. Enough?

There is no shorthand or substitute for truly putting in the time and energy in getting to know your people. Even when you do, you'll find that you've barely scratched the surface. I've been married nineteen years and my wife is still a mystery to me. Every time I think I've got her figured out, she does something that I didn't see coming. That's the beauty of people—and the inherent difficulty. We're a complex and unpredictable lot. We readily embrace competing ideologies yet believe ourselves to be consistent. I don't blame companies for seeking ways to make the oft messy people side less confusing and more predictable. Who wouldn't want that? But come on; how scientifically naïve do you have to be to think you can actually get that kind of accuracy and insight from a thirty-item questionnaire?

CLASHING COPING STYLES VERSUS PERSONALITY CONFLICTS

When we talk about personality conflicts there is a much simpler explana-tion than the mysterious forces that are supposedly unearthed by person-ality tests. Most interpersonal conflict can be explained by the different ways that each of us has learned to cope with life's challenges—and how these differences, at times, clash.

As someone who is trained in cognitive behavioral methods, I don't put much stock in delving heavily into family histories to help people make positive changes in their lives. But I do find it very useful for clients to understand where they learned certain behaviors so we can meaningfully explore ways to modify them. Since these learned behaviors become our default positions in times of trouble, like a needle on a record, they are often the place where we get stuck over and over again. But we can learn to identify and then substitute one skill set for another, and thus push the needle out of its rut.

Most challenges that we face cue up anxiety for us. People learn all sorts of ways to try to turn off unpleasant fight-or-flight-induced sensations. Some are more effective than others, but they all work to varying degrees. Let me give you some examples.

Do you have someone who drives you crazy because he constantly seeks reassurance or acts like he doesn't know how to do something that he knows very well how to do? Many people who act this way came from extremely large families where some form of abuse or neglect was rampant. As kids, they were often the glue that held their families together by taking on adult roles like cooking, cleaning, getting siblings dressed for school, or breaking up their parents' fights—all in an effort to ameliorate the chaos swirling around them. Even though what was asked of them was often well beyond their capabilities, they somehow managed to pull off the impossible. But being kids, they often found themselves feeling completely overwhelmed. Amid these feelings, worry and its close cousin, self-doubt, crept in. And, whenever they are under pressure, these feelings crop up again, even as adults. The amygdala makes sure of this. So even though they are now fully capable of handling most situations, they believe themselves to be inadequate and tend to overamplify the significance of inconsequential gaps in their knowledge base. They therefore can become easily paralyzed over details that to the rest of their teammates seem inexplicably trivial. That's why people with this style constantly seek out reassurance. They are afraid of the catastrophic results that could occur if they happened to make an error—even though in the workplace (unlike their lives at home) most mistakes are rarely catastrophic, and are, in fact, great learning opportunities.

A different person, raised in the same set of circumstances, may have acquired, through trial and error, an entirely different set of coping responses. He may have learned to become adept at escape. He either fled the overwhelming situation or found ways to get others to do his bidding for him by feigning helplessness or acting much younger than he truly was. It's easy to belittle this approach to life, but remember, we're talking about survival here. Unfortunately, this is the least effective survival strategy in the workplace. Ducking assignments or responsibilities, avoiding anxiety-provoking tasks, or inducing others to pick up unfinished work will not trigger the nurturing response in coworkers as it might have with family members. Instead, it triggers anger and resentment—and may bring on the very consequence (termination) that the person was seeking to avoid.

Yet another person may have learned ways of behaving that are on the opposite end of the coping continuum. Again, through trial and error, she may have learned to discount the world of feelings, including her own, primarily because she had to. In the brief moments that she thought about how badly she felt, she became paralyzed, so she learned to shut this type of thinking off. Instead, she took on a veneer of toughness, paired with decisive action, until these responses eventually came to define her. At work, she is a master coper and often appears fearless in most situations. *Appears* is the keyword, as she is often scared out of her wits, but has become so adept at shoving these feelings down that they are entirely unrecognizable—even to herself. Unfortunately, this tendency often renders her incapable of understanding the feelings of others. She often comes off as cold and lacking empathy. Seeing the honest expression of feelings (vulnerability) as a weakness, she tends to rely on bullying to get her way. Coincidentally, this type of coping style fits well with the blow-and-go world of construction, and she will do fairly well as a general foreman, superintendent, or even project manager (PM). But this is where her career will hit a brick wall. Her inability to read people, that is, consider their emotions, and understand their frustrations, so she can resolve disputes and effectively negotiate contracts and claims, prevents her from moving into higher positions—even though her intellect may be more than on par with that of the project executives and vice presidents above her. In short, bullying may work with vendors, laborers, or subcontractors, but it is ineffective with owners, community leaders, government agencies, or other high-ranking managers.

This is just a small sampling of the different coping styles that people can acquire. So, here is the rub: What happens when we put people with different coping styles together in the same trailer? The problem for everyone is that their inner experience is what becomes true and right for them. In fact, acting counter to it feels wrong. By extension, people believe that what is right and true for them in stressful situations is what should be right and true for everyone else. If we tend to be decisive and shut down feelings when under stress, we believe that others should do the same. And when they don't, we wonder (sometimes loudly) what the heck is wrong with them?

This is where skillful, mature leadership comes into play. By using the tools in this book, slowing things down enough to get a good read on what is freaking everyone out, and making room for everyone on the team to get heard, you will enable what is true and right to emerge naturally.

For those of you thinking that I have been guilty of the same method of profiling that I have loudly decried personality testing for, let me hasten to add that there is a vast difference between taking the time to fully assess what someone's coping style is and then teaching him or her how to substitute a more useful attitude and behavior, and merely giving him or her a questionnaire, sticking a label on him or her, and assuming that this is who he or she is and always will be. Coping skills are just that—skills. They don't define who a person is.

To me, the appeal of personality testing—the promise of a magic pill of understanding—underscores its inherent flaw. Since people can't change their personalities, looking at a personality profile is pretty much of a dead-end street. But people can change their coping styles, or learn ways to adapt to the coping styles of others, and this is where our energy should be focused.

Since you don't have the luxury or training to be a psychotherapist, and no personality test is going to give you what you are hoping for, let me suggest a few ways to help you gain a true understanding of the people that you work with.

Start by taking your management team out to dinner and asking everyone to answer the following questions:

Personal:
1. Where were you born and raised?
2. How did you get into the industry?
3. What's most important to you outside of work?
4. What are your hobbies and interests?

On the job:
1. What is your biggest pet peeve?
2. What makes you lose it?

3. What's your favorite form of communication (email, face to face, phone, letter, telegram)? What's your least favorite?
4. How would I know that you are overloaded?
5. How do you usually handle stress (yell, shut down, get quiet, talk nonstop, etc.)?
6. How comfortable are you asking for help?
7. What is your idea of teamwork?
8. In what ways are you a good teammate?
9. What do you think you need to work on as a teammate?

I guarantee that if you and your management team take the time to have an honest dialogue around even four or five of these questions, by the end of the night, you will gain much greater insight into how each of you "tick" than you would by answering some convoluted questionnaire and trying to interpret its result. When you are done, you'll also be able to put your stamp on what you hope to see from them as teammates in the future:

> It's pretty clear that we all have things that we like and things that irritate us, and that we all have very different ways of handling things when we're upset. Some of us shut down, some of us go for a walk, some of us yell, some of us seek out other opinions, and some of us write nasty emails. That just makes us all human. But, let me put out there what I'd like to see us do for the life of this project. I'd like people to be able to say what they need and admit to what they don't know without being made to feel bad about it. When we get pissed off with one another—and we will—I want us to be up front about it, rather than talking to everybody else except the person we're upset with.
>
> And I'm going to tell you something that you've all probably figured out already—I'm not perfect. I'm going to make mistakes. But when I do, I'm going to try hard to be up front with you about it. All I want is for you all to do the same. And if you see that I've made a mistake, don't be shy about saying something about it. I may not like hearing it right away, but I promise that I will hear you.
>
> As I see it, we're all different, but we all need to rely on each other. If we don't trust each other, the people below us will pick up on it and this will infect the whole team. So, I want to make sure that we, as managers, spend some time checking in with each other, at least once a week, in terms of how we are doing as a management team.

Now, isn't this somebody that you'd like to work for? Isn't this somebody you could be productive with? How do you know? Because you can feel it—and no personality profile is ever going to give you that!

Conclusion: The Human Condition

Now that you have come to the end of this little tome, you are probably expecting some witty and pithy exhortation about how after reading this book, everything in your leadership life is going to be wonderful and trouble-free. Unfortunately, I can't do that. But why would you expect anything else? As Nietzsche said,

> Examine the lives of the best and most fruitful people and ask yourselves whether a tree that is supposed to grow to a proud height can dispense with bad weather and storms; whether misfortune and external resistance, some kinds of hatred, jealousy, stubbornness, mistrust, hardness, avarice, and violence do not belong among the favorable conditions without which any great growth even virtue is scarcely possible. (De Botten, 2000, 215)

What I can guarantee you is this: as long as you are in the role of leading people, you can count on being disappointed—a lot.

The person whose numerous complaints you busted your butt to accommodate will unexpectedly quit—and probably at the most inopportune time. That up-and-coming manager-to-be, who sounded so humble and sincere, and a little too good to be true, will turn out to be. And there will come that day when the one person who you always thought would have your back won't. Or maybe it will happen in reverse. The person whose back you always promised to watch will one day do something so inexplicably foolish that it will be impossible, in good conscience, to stand by him. Such are the inescapable foibles of the human condition. Try as we all might, we are all imperfect, and at times, given the wrong set of circumstances, we can all fall woefully short of the ideals we set out for ourselves. As Seneca so wisely said,

> Nothing, whether public or private, is stable; the destinies of men,
> No less than those of cities are a whirl …
> Mortal have you been born, to mortals have you given birth.
> Reckon on everything, expect everything. (De Botten, 2000, 91)

But what I can also reassure you of is if you remain pure of heart and fervently resist the black-hole-like pull toward cynicism, there will also be days of unfathomable satisfaction: Like when that youngster with tons of potential, who you rode relentlessly in the hope of breaking him of his arrogance, years later calls out of the blue to ask your forgiveness for all the crap he put you through. Or when the electrician who you've been bugging for years to put down his tool belt and join the company as a general foreman unexpectedly comes into your office and, with a wide grin, says, "Okay, you win—I'm in." Or that day when you walk past the conference room and see that young lady, the one who you thought was way too timid to ever make it in this industry, standing in front of a bunch of burly men twice her size and age, confidently giving an overview of the updated schedule, as they listen to her attentively. In these moments, you will say to yourself, "Wow, I can't believe what a really cool job I have." And this is also part of the human condition.

At times, you'll wonder why these moments of grace are so elusive—why it is that you can't replicate or sustain them with more regularity. But as said many times, and in many different ways throughout the course of this book, people are complicated. Anyone who tells you otherwise is a two-bit fraud.

People arrive at your project with a plethora of life experiences. They come with differing IQs, family histories, and religious beliefs. Some have traveled the world, while others have never left the neighborhood. Some believe that working hard is everything, while others believe that playing is all that matters. Some are parents, or take care of parents, and some have no deep connections to family life at all. Some drink and party until the wee hours of the morning, while others are homebodies who have never touched a drop of alcohol in their lives. Some dream of being somewhere else, while others are content to be right where they are. Some dream of big things for their careers, while others are satisfied with the particular niche they have carved out for themselves. Some are avid learners, taking in all the information that they can, while others believe that they already know all that there is to know. Some feel that it is better to express their angst directly, while others think it is wiser to squash it down and suffer in silence. The combinations and permutations are seemingly limitless. On top of this, each individual comes with his or her unique set of beliefs about what is right and what is wrong—and is constantly judging his or her world and others against this internal standard.

Every team, in terms of its interpersonal dynamics, is like a ship caught in rough seas, and each is in need of a consistent leader who is willing to man the rudder amid the threatening currents. No one is as good as he thinks he is, or as bad as others perceive him to be. The true art of leadership is not to fight the currents or to bemoan one's task, but to steer a steady course between perception and reality. And despite what you may hear to the contrary, each leader must discover the best way to assume the helm for himself or herself.

So the next time someone tells you that he knows all the secrets of how to be the perfect leader (including me), tell him to take a long leap off a short pier—because there is no such thing.

What works well with one person or one team could very well blow up in your face with the next. All you can do is keep your eyes, ears, and heart open, maintain what you believe to be right, true, and good, and steer a consistent course. Everything else is more or less gravy.

The fact is leadership done well is a paradox. It is a well-intentioned act of deception. In the same way that religion allows us to rise above life's indifference, banality, and suffering so that we may see something beautiful in it all, so too does effective leadership induce a temporary willful suspension of belief. It allows those under us to put aside their selfishness and all the little self-justified treacheries that are waged under the seemingly benign cloak of ambition and, for a time, produce something that is well beyond the capabilities of any one individual. It is that feeling that is captured in the moment when you look up at the soon-to-be-completed project and ponder: How could something so massive and so otherworldly—something that breathes and flexes and pulses with life—have been made by such small crude hands? And as impossible as it seems, you know it was, and that you played some small, yet significant part in directing those hands.

If you have as much gray around your temples as I have, you'll have long ago discovered the foolishness of holding on to absolutes. But what does this mean as a construction leader? It means that there will be a time to yell and a time to calm down; a time to show the way and a time to push forward; a time to encourage and a time to say "hell no"; a time to be unyielding and a time to give generously. For each situation, and each person that we work with, there is a response that is good and true that lies within us. It's just that sometimes we need to take a deep breath, in order to see that what we really need to do is give ourselves over to the moment and allow ourselves the time to find it.

So maybe, if you're a yeller, from now on, you'll give yourself a couple of more minutes to decide if that's what you really want to do, and if you do, maybe you'll take it down a notch or two to make sure that it is the message, not the messenger, that is being heard. Or if your tendency is to jump to conclusions, maybe you'll ask some additional key questions to make sure that you've gotten all of your facts straight first. Or if you tend to ponder and avoid any and all forms of conflict, maybe today is the day that you'll decide to sit down and engage that person who has been detrimentally ruffling the team's feathers and set him straight on what you think is really important in terms of teamwork. Or if (egad!) you are the ultimate control freak, maybe, just maybe, you'll decide that today is the day that you'll make the commitment to find something—just one small thing—to let go of, and do the same each day forward. Or maybe, because you've simply grown tired of enabling the same broken systems to limp along day in and day out, today is the day that you teach your boss the five why's, and fix the root causes that have been plaguing your projects for as long as you can remember.

And here lies yet another paradox: Believe it or not, by engaging in these simple beautiful acts, you will convey an essential truth to your team—that by caring about *how* they do their work, by extension, you care about them. If you think deeper about all this work-life balance talk that is going on in the industry right now, isn't this the essence of Lean culture and Lean thinking? If we can be more thoughtful about how a job is organized; if we can increase our efficiency in terms of preparation, information flow, and coordination; and if, by these acts, we can, in turn, reduce the amount of wasted time and effort that they expend, then not only will our people feel less stressed about what they have to do on the job, but also they will also be able to go home earlier and spend more time with their families—a much better alternative to spending needless hours cooped up behind a computer screen inside of some cramped and drafty trailer.

On those dark days when you feel like nothing you do is ever right, try to remember that you are taking part in something noble. Being a good leader means taking your place alongside those who, in the heat of battle, remained not only strong, but also selfless. History teaches us that being a leader is a lesson not in bravado, but in humility. People aren't going to follow you simply because you have a title or because you tell them to do so forcefully. It is about helping others to connect with something that is truly bigger than themselves, while at the same time knowing that when you look inside yourself, the qualities you are trying to attain will always remain tantalizingly out of reach. But it is in the striving, not in the possessing, that makes a leader truly great.

> A gentleman leader has nine aims: To see clearly; to understand what he hears; to be warm in manner, dignified in bearing, faithful of speech, painstaking at work; to ask when in doubt; in anger to think of the difficulties anger may bring; in sight of gain to remember right…. Effective leaders are virtuous leaders. Wisdom, benevolence and courage; these are the three universal virtues. Some practice them with ease of nature; some for the sake of their own advantage; and some by dint of great effort.

> **The Analects of Confucian (Ames, 1998)**

Confucious uttered these words more than two thousand years ago, but they still hold true today. So please, take the time to relish the rare opportunity that you have to impact people's lives, even if you find that it takes great effort. And on those days (and there will be many), when everything feels as dark as a New England winter, recall what Mother Teresa was fond of saying: "God asks us to be faithful, not successful; success is God's responsibility."

So, have faith my good and noble friends. Be relentless in your quest. Once you realize that there is no way to control the complexities of the human condition other than to accept them fully orchard even do—and commit to doing your best to utilize yourself as a true leader each and every day—you will be well on your way to building the culture that you seek. To paraphrase a philosopher much wiser than myself, "Can you build a Lean culture? Yes you can!" Perhaps this final quote, often cited by Nelson-Mandela, says it best:

> Our deepest fear is not that we are inadequate. Our deepest fear is that we are powerful beyond measure. It is our light, not our darkness, that most frightens us…. your playing small doesn't serve the world. There's nothing enlightened about shrinking so that others won't feel insecure… as we let our own light shine, we unconsciously give other people permission to do the same. As we are liberted from our own fear, our presence automatically liberates others."

> **Marianne Williamson (1992)**

Bibliography

AECbytes. www.AECbytes.com.

Ames, R.T. and Rosemott, H. 1998. *The analects of Confucius: A philosophical translation*. New York: Random House.

Arbinger Institute. 2000. *Leadership and self-deception*. San Francisco: Berrett-Koehler.

Aunger, R. 2002. *The electric meme: A new theory of how we think*. New York: The Free Press.

Bourdain, A. 2000. *Kitchen confidential*. New York: Bloomsbury USA.

Bourdain, A. 2001. *A cook's tour*. New York: Bloomsbury USA.

BrainyQuote.com. Xplore Inc, 2010 *Fyodor Dostoevsky*. (24 October. 2010.) http://www.brainyquote.com/quotes/quotes/f/fyodordost124427.html.

Burley-Allan, M. 1995. *Listening: The forgotten skill. A self teaching guide*. New York: Wiley.

Csikzentmihalyi, M. 1997, *Finding flow: The psychology of engagement with everyday life*. New York: Basic Books.

Damasio, A. 1994. *Descartes' error*. New York: G.P. Putnam.

Daniels, A. 1989. *Performance management*. Tucker, GA: Performance Management Publications.

De Botton, A. 2000. *The consolations of philosophy*. New York: Vintage International.

De Botton, A. 2004. *Status anxiety*. London: Penguin Group.

Elbing, A., and Elbing, C. 1991. *Do aggressive managers really get high performance?* Sarasota, FL: Scott Foresman Professional Books.

en.wikipedia.org/wiki/tenerife_airport_disaster.

Frankl, V. E. 1959. *Man's search for meaning*. New York: Simon & Schuster.

Gambrill, E., and Gibbs, L. 2009. *Critical thinking for helping professionals*. Oxford: Oxford University Press,

Goldratt, E. 2004. *The goal: A process of ongoing improvement*, Great Barrington: MA: North River Press.

Goleman, D. 2005. *Emotional intelligence: Why it can matter more than I.Q.* 10th anniv. ed. New York: Bantam Books.

Krause, S. 2003. *Aircraft safety: Accident investigations, analysis, and applications*. New York: McGraw-Hill.

Lencioni, P. 2002. *The five dysfunctions of a team*. New York: Jossey-Bass.

Liker, J. K. 2004. *The Toyota way*. New York: McGraw Hill.

Macomber, H. 2004. *Securing reliable promises on projects*. Weblog halmacomber.com.

Martin, R. 2002. *The responsibility virus*. New York: Basic Books.

Meichenbaum, D., and Turk, D. C. 1987. *Facilitating treatment adherence*. New York: Plenum Press.

Moser, K. 2009. *Group think. Project teams find integrated delivery very helpful*. www.comstockbusiness.com.

Nee, P. A. 1996. *ISO 9000 in construction*. New York: John Wiley & Sons.

Paul, A. M. 2005. *The cult of personality testing*. New York: Free Press.

Pinker, S. 2002. *The blank slate*. New York: Viking.

Sapolsky, R. M. 1994. *Why zebras don't get ulcers*. New York: W. H. Freeman.

Souden, D. 1997. *Stonehenge revealed*. New York: Facts On File Books.

Teicholz, P. 2004. *Labor productivity declines in the construction industry: Causes and remedies*. AECbytes Viewpoints #4 (April 4, 2004) http://www.aecbytes.com/viewpoint/2004/issue_4.html

Tjosvold, D., and Tjosvold, M. M. 1995. *Psychology for leaders*. New York: John Wiley & Sons.

Vaillant, G. E. 1977. *Adaption to life*. Boston: Little, Brown and Company.

Williamson, M. 1992. *A return to love*. New York: HarperCollins.

Index